The Economic Challenge for Europe

The Economic Challenge for Europe

Adapting to Innovation Based Growth

Edited by

Jan Fagerberg

Professor of Economics, University of Oslo, and Senior Research Fellow, Norwegian Institute of International Affairs, Norway

Paolo Guerrieri

Professor of Economics, University of Rome 'La Sapienza', Italy

Bart Verspagen

Professor of Economics of Technological Change, ECIS, Eindhoven University of Technology, and MERIT, Maastricht University, The Netherlands

Edward Elgar

Cheltenham, UK • Northampton, MA, USA

Published by
Edward Elgar Publishing Limited
Glensanda House
Montpellier Parade
Cheltenham
Glos GL50 1UA
UK

Edward Elgar Publishing, Inc.
136 West Street
Suite 202
Northampton
Massachusetts 01060
USA

A catalogue record for this book is available from the British Library

Library of Congress Cataloguing in Publication Data

The economic challenge for Europe : adapting to innovation based growth / edited by Jan Fagerberg, Paolo Guerrieri, Bart Verspagen.

"This book originates from a research project, 'Technology, Economic Integration and Social Cohesion' (TEIS), carried out with support from the European Union's Targeted Socio-Economic Research Programme (TSER)"—Pref.

1. Technological innovations—Economic aspects—Europe. 2. Competition—Europe. 3. Europe—Economic policy. I. Fagerberg, Jan. II. Guerrieri, Paolo, 1947– . III. Verspagen, Bart.

HC240.9.T4E27 1999

338'.064'094—dc21 99–14855

CIP

ISBN 1 84064 124 X (cased)

Printed in the United Kingdom at the University Press, Cambridge

Contents

Figures

Tables

Contributors

Iain Begg, South Bank University, London, United Kingdom.
Aadne Cappelen, Statistics Norway, Oslo, Norway.
Bent Dalum, IKE Group – Department of Business Studies, Aalborg University, Aalborg, Denmark.
Jan Fagerberg, University of Oslo, and Norwegian Institute of International Affairs (NUPI), Oslo, Norway.
Chris Freeman, Science Policy Research Unit, University of Sussex, Brighton, United Kingdom.
Paolo Guerrieri, University of Rome 'La Sapienza', Rome, Italy.
Heather Grabbe, European Programme, The Royal Institute of International Affairs, London, United Kingdom.
Magnus Holmén, Industrial Dynamics, Chalmers University of Technology, Gothenburg, Sweden.
Kirsty Hughes, European Programme, The Royal Institute of International Affairs, London, United Kingdom.
Staffan Jacobsson, Industrial Dynamics, Chalmers University of Technology, Gothenburg, Sweden.
Marion Jansen, Maxwell Stamp plc, London, United Kingdom.
Michael Landesmann, The Vienna Institute of International Economic Studies and Department of Economics, Johannes Kepler University, Linz Auhof, Austria.
Per Botolf Maurseth, Norwegian Institute of International Affairs (NUPI), Oslo, Norway.
Mario Pianta, CNR – ISRDS, Rome, Italy.
Mette Praest, IKE Group – Department of Business Studies, Aalborg University, Aalborg, Denmark.
Annika Rickne, Industrial Dynamics, Chalmers University of Technology, Gothenburg, Sweden.
Roberto Simonetti, Faculty of Social Sciences, Open University, Walton Hall, Milton Keynes, United Kingdom.
Nick von Tunzelmann, Science Policy Research Unit, University of Sussex, Brighton, United Kingdom.
Bart Verspagen, ECIS, Eindhoven University of Technology, and MERIT, Maastricht University, The Netherlands.

Gert Villumsen, IKE Group – Department of Business Studies, Aalborg University, Aalborg, Denmark.
Marco Vivarelli, Faculty of Economics, University of Piacenza, Piacenza, Italy.

Preface

This book originates from a research project, 'Technology, Economic Integration and Social Cohesion' (TEIS), carried out with support from the European Union's Targeted Socio-Economic Research programme (TSER). It is the first of a series of books from the project, and applies a long-run perspective to the study of European growth and competitiveness.

The TEIS project has been an unusual one in several respects. First, it has had a wide thematic focus, ranging from technology and globalization to labour markets and social cohesion. Second, based on this wide focus, it has aimed at developing a holistic perspective on Europe's current situation and future. Third, it has been an unusually large project, bringing together more than 45 researchers from nine countries and 18 research institutions. Fourth, there has been a systematic attempt to stimulate collaborative work between researchers from different countries and institutions, and the lists of contents and contributors in this volume testify to the success of this aim. Hence, it is a truly collective European research effort that is reported in these pages. Without the generous support from the EU's TSER programme this would not have been possible. The Norwegian Research Council provided valuable support in the initial stage. The project has also benefited from co-financing from many of the institutions taking part in the project and from national sources.

As editors of this volume we would like to thank all the people who contributed to the process that led to its publication. This includes a great number of people who have participated in discussions at workshops and conferences, and administrative support personnel. Special thanks to Vibeke Sand and Eilert Struksnes at NUPI for their assistance in the final editing process.

Jan Fagerberg, Paolo Guerrieri and Bart Verspagen
Oslo/Rome/Maastricht, 1 December 1998

1. Europe – a long view

Jan Fagerberg, Paolo Guerrieri and Bart Verspagen

1 INTRODUCTION

This book is about Europe's economic performance. In contrast to many contemporary writings on the subject we deliberately take a long-run perspective. Nowadays many people seem to take it for granted that what mainly matters for European growth is how fiscal and monetary policy is handled. However, there is not much scholarly support for that position. For instance, Levine and Renelt (1992), in a very thorough investigation based on data for a large number of countries, found that there is no robust correlation between the fiscal and monetary policy stance and growth. Hence, although fiscal and monetary policies do have a say, particularly in the short to medium run, the real sources of differences in long-run growth will have to be found elsewhere.

For a long time, economists used to focus on differences in the supply of capital and 'land' (that is, national resources) to explain differences in growth performance. However, it is now commonly accepted that the ultimate source of economic growth is to be found in ideas, or human creativity, including the ability to apply new ideas to practical uses in an efficient manner. Hence, this book concentrates on questions related to innovation and diffusion of technology, particularly in the European context, and the interaction between innovation diffusion and long-run growth/competitiveness.

Admittedly, this shift in perspective, important as it may seem, does in itself not provide much guidance for policy. All too often one observes policy-makers complaining about the level of innovation or R&D, without having very concrete ideas of why this is so, or what to do about it. The point is that without a clear understanding of what shapes innovation processes in Europe, and what the likely outcomes are, it becomes very difficult to design policy in this area. In many ways the situation may be compared to the one that prevailed more than fifty years ago, when the so-called Keynesian revolution in economic thinking made old policies obsolete, while at the same time the design of new policies was hampered by lack of relevant methods, data, etc. The purpose of this book is to fill some of the more important gaps in our current knowledge, and in this way contribute to the discussion of the policy alternatives in front of us.

As an introduction to the analysis, the next sections present some 'stylized facts' on Europe's long-run growth and trade performance. On this basis we outline the main issues to be analysed in the coming chapters.

2 A LONG VIEW

Today many policy-makers in Europe express concerns about Europe's economic performance compared to other global players. In terms of productivity, as measured by GDP per capita, Europe seems to be surpassed not only by the USA, but also by Japan and – if the trends from the past decades prevail in the next century – by a number of other Asian economies as well.

How did this happen? One or two centuries ago, the situation was entirely different. During most of the nineteenth century the UK was the leading capitalist country in the world, with a GDP per capita about 50 per cent above the average of other leading capitalist countries (Table 1.1). This lead was, among other things, related to the process of economic, social and institutional change that had taken place in Britain for quite some time already, the so-called Industrial Revolution (von Tunzelmann, 1995).

However, during the second half of the century, the USA started to catch up with the UK and eventually – around 1910 – surpassed it. In retrospect it becomes clear that US growth was based on the development of a new technological system, based not so much on new products as on a new way to organize production and distribution (taylorism, fordism, etc.). The large productivity gains were secured through the development of large-scale production and distribution systems well suited for the large, fast-growing and relatively homogeneous American market (Chandler, 1990; Nelson and Wright, 1992). The developments of standards and interchangeable parts were important ingredients in this process.

That Europe initially failed to take advantage of these innovations is perhaps not so difficult to explain. For example, one main difference between the USA and Europe in the first half of this century relates to the size of markets. The European markets were smaller and less homogeneous. Hence, it is not obvious that US methods, if applied to European conditions in this period, would have yielded superior results. This is what Abramovitz (1994) has dubbed a lack of 'technological congruence'. Two world wars and an intermediate period of protectionism and slow growth added to these problems (Abramovitz, 1994). Hence, the US lead increased even further and peaked around 1950, when GDP per capita in the USA was about twice the European level.

Table 1.1 GDP per capita in the industrialized world, 1820–1994 (100 1990 US$ per head)

	1820	1870	1910	1950	1970	1980	1990	1994	Growth 1820–1950	Growth 1950–94(90)
USA	1.3	2.5	5.0	9.6	14.9	18.3	21.9	22.6	1.5	2.0
Japan	0.7	0.7	1.3	1.9	9.4	13.1	18.5	19.5	0.8	5.3
Germany	1.1	1.9	3.5	4.3	11.9	15.4	18.7	19.1	1.0	3.4
France	1.2	1.9	2.9	5.2	11.6	15.0	17.8	18.0	1.1	2.8
Italy	1.1	1.5	2.3	3.4	9.5	13.1	16.0	16.4	0.9	3.6
United Kingdom	1.8	3.3	4.7	6.8	10.7	12.8	16.3	16.4	1.0	2.0
Canada	0.9	1.6	3.9	7.0	11.8	16.3	19.6	18.4	1.6	2.2
Belgium	1.3	2.6	4.0	5.3	10.4	14.0	16.8	17.2	1.1	2.7
Netherlands	1.6	2.6	3.7	5.9	11.7	14.3	16.6	17.2	1.0	2.5
Korea				0.9	2.2	4.1	9.0			5.8
Taiwan			1.0	0.9	2.7	5.6	10.3			6.0
Hong Kong				2.0	5.3	10.0	17.1			5.4
Europe (6)										
Mean	1.3	2.3	3.5	5.2	11.0	14.1	17.0	17.4	1.0	2.8
Coeff. of var.	0.18	0.26	0.22	0.21	0.08	0.07	0.05	0.05		
All countries except Asian NICs										
Mean	1.2	2.1	3.5	5.5	11.3	14.7	18.0	18.3	1.2	2.7
Coeff. of var.	0.25	0.35	0.32	0.38	0.14	0.11	0.10	0.10		
All countries										
Mean				4.4	9.3	12.7	16.5			3.3
Coeff. of var.				0.59	0.4	0.32	0.21			

Notes:
1. For Korea, Taiwan and Hong Kong, the final period is 1950–90, not 1950–94.
2. Europe (6) is the six European countries in the table.

Source: Calculations based on Maddison (1996).

While the period between 1820 and 1950 was one of divergence in economic performance between leading capitalist countries, the following period has generally been one of convergence, as least as far as variables such as income or GDP per capita are concerned. The productivity gap between the USA and Europe has been significantly reduced (cut by one half). Arguably, this reduction, most of which occurred during the 1950s and 1960s, was related to the potential for rapid productivity advance in Europe through imitation (of superior US technology). European production and exports in industries such as cars, domestic electrical equipment, electronics, etc. grew rapidly from the 1950s onwards. The gradual reduction of barriers to trade within Europe from the 1950s onwards has generally been regarded as an important contributing factor to this process, as has the general rise in incomes and living standards (Abramovitz, 1994; Maddison, 1982, 1991).

European countries were not alone, however, in exploiting the window of opportunity given by superior US technology. From the 1950s onwards Japan, later joined by other Asian economies, aggressively targeted the very same industries as those that had grown rapidly in Europe (Johnson, 1982; Wade, 1990). Initially this did not give much reason for concern among policy-makers or industrialists, but during the 1970s and 1980s it became evident that Japanese suppliers outperformed their European and US competitors in many cases, and that this could not be explained solely by low wages. It became clear that the Japanese, as the Americans before them, had made important innovations in the organization of production, innovations that have led to both increased quality and higher productivity (von Tunzelmann, 1995), and which US and European competitors – despite serious efforts – have not yet managed to imitate to the extent that they would have wished. This is partly explained by the fact that process innovations generally are of a more 'tacit' character than product innovations, and therefore more difficult to copy.

While Europe, Japan and other countries started to catch up in many typical 'American way of life' products, US industry leaped forward in another area; science-based industry. Before the First World War – and arguably in the interwar period as well – Europe, and Germany in particular, was at the forefront in this area. In fact, science-based industry, characterized by high R&D investments, R&D departments and close interaction between industry, research institutes and universities, was largely a German invention. However, at the beginning of the twentieth century the US business community started to catch up in this area (Nelson and Wright, 1992; Mowery and Rosenberg, 1993). Technical universities and business schools were founded, often in close interaction with industry. This drive towards a greater reliance on science and R&D was much strengthened during the Second World War and the 'cold war' that followed due to massive public investments in this area. As a consequence, the leadership passed to the USA.

Gradually, however, European countries and Japan started to devote more resources to science and R&D. By the mid-1970s many of these countries used a larger share of their available resources on civil R&D than did the USA (Table 1.2). Hence the US lead started to be questioned in this area as well. Following the Japanese example, some of the Asian NICs started to invest massively in R&D from the 1970s onwards. As we are going to explore in more detail later, these changes have had a major impact on the structure of science-based industry worldwide.

Table 1.2 Estimates of non-defence R&D as a percentage of GDP (selected years)

	1963	1973	1981	1993
USA	1.5	1.7	1.9	2.1
Japan	1.4	1.9	2.4	2.8
Germany	1.3	2.0	2.4	2.3
France	1.2	1.5	1.7	2.1
Italy	0.7	0.9	1.0	1.3
United Kingdom	1.5	1.6	1.9	1.8
Canada	1.0	1.1	1.2	1.6
Belgium	1.0	1.4	1.4	1.6
Netherlands	2.0	2.0	1.9	2.0
Sweden	0.9	1.4	2.0	3.0
Switzerland	2.5	2.3	2.3	2.6
Korea	0.3	0.3	0.8	2.3
Taiwan	0.4		0.7	1.8
Europe (6)				
Mean	1.3	1.6	1.7	1.9
Coeff. of var.	0.3	0.2	0.2	0.2
All countries except Asian NICs				
Mean	1.4	1.6	1.8	2.1
Coeff. of var.	0.4	0.2	0.2	0.2
All countries				
Mean	1.2	1.4	1.7	2.1
Coeff. of var.	0.5	0.4	0.3	0.2

Note. Estimates of civil R&D are available for most OECD countries for selected years. These estimates show that for all but a few countries (notably the USA, the UK, France and Sweden) differences between civil and total R&D are small. Hence, for some countries and years, total R&D is used instead. Since data are not always available annually, the data reported here will in some cases be based on information from a preceding and/or following year. For definition of Europe (6), see Table 1.1.

Source: Calculations/estimates based on data from OECD Science and Technology Indicators Database and UNESCO/national sources (Korea, Taiwan).

During the 1980s European catch-up *vis-à-vis* the USA came to a halt, while Japan continued to increase productivity at a faster rate than both the USA and Europe. As a consequence Japan now enjoys a higher level of GDP per capita than Europe. The Asian NICs (Hong Kong, Korea and Taiwan) have – at least until very recently – continued to experience very rapid productivity growth, that is, catch-up relative to the USA and other countries.

3 STRUCTURAL CHANGE IN THE WORLD ECONOMY

The process of long-run economic growth described in the previous section was one in which structural change played an important role. The take-over of technological and economic leadership by the USA at the beginning of the twentieth century stimulated the diffusion of manufacturing products and the system of mass production that was at the heart of the American economy, throughout the world economy. From this perspective, Europe's and Japan's catch-up process during the period 1950–85 can be seen as processes of (further) industrialization.

Table 1.3 illustrates the structural changes that took place over the postwar period. It is clear that in 1950, the USA and Europe as a whole (although this mainly reflects the data for the larger European countries) were clearly already industrialized societies, with manufacturing shares in total GDP well above 20 per cent, and agriculture shares around 5 per cent. However, Japan and the other two Asian countries for which we have data, (South) Korea and Taiwan, as well as some of the individual European countries, such as Italy, Spain, Denmark and the Netherlands, still had fairly large shares of agriculture in GDP.

In the period after 1950, Japan quickly increased its share of manufacturing in GDP, as did most of the European countries. In fact, in all European countries, except Denmark and the UK, the share of manufacturing in GDP increased monotonically over the period 1950–70. In the USA, on the other hand, the share of manufacturing remained constant after 1950. For Europe as a whole, manufacturing increased its share up to the 1970s, after which a (weak) trend of de-industrialization set in. However, in Japan the manufacturing sector kept increasing its share until 1990. The two Asian NICs, Korea and Taiwan, also showed a strong process of industrialization.

Agriculture became a smaller share of GDP in all countries after 1950. The fall was largest in Japan as well as the above-mentioned European countries, although also in the USA and the other European countries, the share of agriculture fell.

Table 1.3 Shares of selected sectors in GDP (per cent), 1990 constant prices

	1950	1960	1970	1980	1990	1995
Agriculture						
(West) Germany	4	3	2	2	2	2
France	9	8	6	4	4	4
UK	2	2	2	2	2	2
Italy	11	9	7	5	4	4
Denmark	10	9	5	5	6	7
Netherlands	4	4	3	3	5	5
Sweden	9	7	4	3	3	3
Spain	17	13	8	6	5	5
Europe*	5	4	3	3	3	3
Japan	16	15	5	3	3	na
USA	6	4	3	2	3	3
Korea	41	38	29	14	8	7
Taiwan	32	25	15	7	4	3
Manufacturing						
(West) Germany	26	34	37	34	31	27
France	18	20	25	25	22	22
UK	27	30	30	26	25	22
Italy	12	14	17	22	23	23
Denmark	19	19	19	20	17	17
Netherlands	13	18	20	17	18	18
Sweden	20	21	26	23	22	25
Spain	12	13	20	26	25	na
Europe*	23	28	30	28	25	24
Japan	13	16	25	27	31	na
USA	22	21	22	22	22	22
Korea	3	6	14	27	34	30
Taiwan	15	16	27	35	34	31
Financial and business services						
(West) Germany	8	9	10	12	14	15
France	13	13	15	18	19	20
UK	12	12	12	16	20	20
Italy	12	11	10	11	11	12
Denmark	10	10	14	16	17	16
Netherlands	11	15	18	18	21	22
Sweden	16	18	16	17	18	19

Table 1.3 continued

	1950	1960	1970	1980	1990	1995
Spain	17	17	18	17	17	na
Europe*	11	12	13	15	18	17
Japan	14	13	12	15	16	na
USA	10	12	13	15	15	14
Korea	12	12	10	11	14	17
Taiwan	10	8	10	14	17	17
Transport and communications						
(West) Germany	6	5	5	6	6	7
France	5	5	5	6	8	8
UK	6	6	7	7	8	8
Italy	5	6	6	7	8	9
Denmark	10	11	9	8	9	11
Netherlands	5	7	7	7	7	9
Sweden	5	5	5	6	8	8
Spain	3	3	4	5	5	na
Europe*	6	6	6	6	7	8
Japan	5	6	8	6	6	na
USA	6	5	5	6	7	8
Korea	1	2	5	8	8	8
Taiwan	4	3	4	6	6	6

Notes:
Values listed as 1950 for Korea are for 1953; values listed as 1960 for Taiwan are for 1961; values listed as 1950 for Taiwan are for 1951 and in current prices; values listed as 1995 for Taiwan are for 1993.
* Europe denotes the weighted average of all European countries in the table.

Source: Calculations based on data in Van Ark (1993), updated with the 1998 version of the OECD ISDB database, data for Korea are from Pilat (1993), and from Korean National Accounts (1995). Data for Taiwan are from Taiwanese National Accounts.

Regarding services, the two services industries listed in Table 1.3, financial and business services, and transport and communications, have been among the most rapidly growing. In the USA, the share of financial and business services in GDP increased by 50 per cent between 1950 and 1980. Also, in Europe, financial and business services increased its share rapidly, but a large part of the growth came later (in the 1980s). The share of transport and communications also increased in both the USA and Europe, but much less. The developments for the Asian NICs to some extent resemble those of Europe, with rapid growth in financial and business services after 1980. In contrast to the other countries, the data for Japan indicate a much more stable pattern for these two service industries (as shares of GDP).

There are two main conclusions from these data. First, the process of catching up to the USA by Europe, Japan and the Asian NICs was one which was accompanied, and to a certain extent driven, by an increase in the manufacturing sector. While in Asia this process of industrialization continued well into the 1990s, Europe diverged into a (weak) de-industrialization path from the 1970s onwards. Second, services, especially financial and business services, increased its share in GDP in all three blocks (though less in Japan than in the two other Asian NICs). Hence, the core of the world economy is thus becoming one in which services play a larger role (as compared to manufacturing).

Europe developed, in this way, a production structure that is relatively similar to the one in the USA, but rather different from that of Japan and the Asian NICs. What are the reasons for – and consequences of – these developments? To what extent are they related to Europe's relatively weak economic performance in recent years? These are some of the questions that we address in the following chapters.

4 EUROPE IN THE GLOBAL MARKET

Technological catch-up (or lack of such) and structural changes are intimately related to trade performance. In fact, one of the striking findings from studies of successful catch-up is that it is associated with both a general improvement in trade performance and a radical change in the composition of trade. For instance, the catch-up of Europe and Japan in scale- and capital-intensive technologies from the 1950s onwards was associated with rapidly increasing export market shares for products embodying these technologies. However, since then the locus of growth within manufacturing has changed from scale- and capital-intensive to science-based industries. How has Europe adapted to these changes in the global market?

As a preliminary assessment of Europe's performance in this area we will, following work by Pavitt (1984, 1988), make use of a taxonomy of industrial sectors in which the industries are allocated to sectors depending on the nature and sources of technological knowledge. The taxonomy that will be used here identifies five types of industries: *agricultural products and raw materials*, *traditional (supplier-dominated)*, *scale-intensive*, *specialized suppliers* and, finally, *science-based* industries. Each type represents a different style of technological learning.

The *agricultural products and raw materials* sector, as defined in this chapter, includes agricultural products, raw materials, fuels and natural resource-intensive industries such as petroleum, refineries, non-ferrous metal basic industries, pulp and paper. In this sector the availability of abundant raw materials strongly influences the location of production and countries' comparative advantages. The *traditional* (*supplier-dominated*) sector, on the other hand, encompasses a number of traditional industries that use low-skill labour intensively, such as textiles, clothing, furniture, leather and shoes, ceramics and simple metal products. Both sectors are net purchasers of process innovations and innovative intermediate inputs from other sectors. Thus, technical change mostly comes from suppliers of machinery, etc. (hence the term 'supplier-dominated'). In both sectors technology is easily accessible, international technology transfers are relatively easy and factor endowments have a major influence on the generation of comparative advantages.

Scale-intensive industries, such as automobiles, consumer electronics, consumer durables, rubber and steel, are typically characterized by large oligopolistic firms with high capital intensity, large economies of scale, high technical or managerial complexity and significant in-house technological accumulation (learning) through design, production engineering and operating experience. Process and product technologies therefore develop incrementally.

Specialized suppliers include most producers of machinery, components and software in mechanical engineering and instruments (such as machinery for specialized industries). These industries are characterized by a high diversification of supply, high 'economies of scope', medium to small firms and a notable capacity for product innovation that diffuses to other sectors as capital inputs (Rosenberg, 1976, 1982; von Hippel, 1988).

Table 1.4 Commodity composition of world trade 1970–95, per cent

	1970	1988	1995	Change 1970–95
Agricultural products and raw materials	36.5	26.6	22.2	–14.2
Traditional	14.9	16.5	16.8	1.9
Scale-intensive	24.7	26.2	25.0	0.3
Specialized suppliers	10.9	10.0	10.5	–0.4
Science-based	9.5	17.5	21.5	12.1
Other	3.6	3.2	3.7	0.1

Source: UN and OECD data from SIE-World Trade Database.

Finally, the so-called '*science-based*' sector includes industries such as fine chemicals, electronic components, computers, telecommunications and aerospace. These industries are all characterized by innovative activities directly linked to high R&D expenditures and academic research. Their product innovations generate broad spillover effects on the whole economic system (OECD, 1992).

Based on earlier work by Guerrieri (1992) we apply this taxonomy to data on international trade (value data) for the period 1970–95. Table 1.4. shows the sectoral composition of world trade in this period. At the beginning of the 1970s the largest sector was agricultural products and raw materials which alone accounted for more than one third of world trade. The second largest sector was the scale-intensive one (one quarter of world trade). The other sectors were all markedly smaller, with the science-based sector being the smallest one (less than one tenth of world trade). In the decades that followed this pattern changed significantly. The share of agriculture and raw materials in world trade declined by more than one third, while the share of science-based industries more than doubled, so that the sectors became roughly comparable in size. The shares of the other sectors remained relatively stable. Hence, the most striking development in the last decades has been the spectacular growth of industries based on organized R&D and exploitation of science.

Table 1.5 reports market shares for four major players in the global economy, Europe (15), the United States, Japan and the Asian NICs, for 1970, 1988 and 1995. In the early 1970s Europe was specialized (that is, having above average market shares) in manufactures rather than agricultural products and raw materials. Compared to the other country groupings, Europe's pattern of specialization in manufacturing was relatively even, without very strong or weak points. In contrast, the USA showed a strong specialization in high-technology industries, particularly science-based industries and specialized suppliers, while Japan reported a similar strength in scale-intensive industry. The Asian NICs showed the strongest specialization of all (in traditional industries).

In the decades that followed, this pattern underwent radical changes. First, Japan and the Asian NICs gained market shares at the expense of the USA and Europe. In particular, the Asian NICs showed a spectacular performance; between 1970 and 1995 their overall market share increased more than five times, from 2.1 to 10.8 per cent of the global market. It is also noteworthy that the rapid growth of Japan and the Asian NICs was accompanied by very rapid structural changes that totally altered the specialization pattern of these countries. In the case of Japan, in spite of overall market share growth, the market shares for traditional and scale-intensive industries actually contracted, while those of specialized suppliers and science-based industry increased rapidly, so that by the end of the period Japan had its major strength in the latter (followed by scale-intensive industry). For the Asian NICs a similar development took place, with a much stronger growth in science-based industry and specialized suppliers than in the traditional area of strength (traditional industries). As a result, at the end of the period the Asian NICs had developed a 'dual' pattern of specialization, with considerable strength in both science-based and traditional industries.

Table 1.5 Market shares 1970–95 (ratio of national exports to world exports, per cent)

		Agricultural products and raw materials*	Traditional industries	Scale-intensive	Specialized suppliers	Science-based	Total
Europe (15)	1970	24.1	57.0	55.7	61.2	48.6	44.6
	1988	30.3	47.6	51.2	56.0	41.3	44.0
	1995	31.6	40.1	47.3	47.6	33.8	39.6
	Change 1970–95	7.5	–16.9	–8.4	–13.6	–14.8	–5.0
US	1970	13.1	7.4	14.5	22.3	29.5	14.8
	1988	13.4	5.2	9.4	12.2	19.8	11.6
	1995	11.0	6.7	10.3	13.7	17.9	11.8
	Change 1970–95	–2.1	–0.7	–4.2	–8.6	–11.6	–3.0
Japan	1970	1.2	9.3	13.8	6.4	7.7	6.7
	1988	1.1	4.1	17.1	15.6	16.7	10.1
	1995	1.4	3.2	12.8	15.7	14.3	9.0
	Change 1970–95	0.2	–6.1	–1.0	9.3	6.6	2.3
Asian NICs	1970	2.0	6.1	1.0	0.8	1.0	2.1
	1988	2.6	14.5	5.6	4.0	9.1	6.7
	1995	3.4	16.2	8.7	8.8	17.8	10.8
	Change 1970–95	1.4	10.1	7.7	8.0	16.8	8.7

Notes:
Europe (15): Austria, Belgium, Denmark, Finland, France, Germany, Greece, Ireland, Italy, Luxembourg, Netherlands, Portugal, Spain, Sweden, United Kingdom.
Asian NICs: Hong Kong, Korea, Singapore, Taiwan.
*For the definition of product-industry groups see Appendix 1.1

Source: UN and OECD data from SIE-World Trade Database.

In contrast to the Asian experience, both Europe and the USA lost overall market shares. Moreover, these losses were generally more manifest in high-technology sectors, particularly science-based industries, than elsewhere. As a consequence, in the 1990s Europe was no longer specialized (had an above average market share) in science-based industry. Europe did a little bit better than the USA in its areas of traditional strength, specialized suppliers and scale-intensive industry, while the USA outperformed Europe in traditional industries. However, the only area in which Europe gained market share was agricultural products and raw materials.

Hence, the picture is very clear. European competitiveness, whether measured through its growth or trade performance, is deteriorating. Its comparative advantage is still based on manufacturing, but the most favourable performance is registered in agricultural products and raw materials. Within manufacturing, scale-intensive industry performed relatively better than the rest, while Europe's position in science-based industry – the fastest growing segment of world trade in recent decades – is being gradually eroded. Moreover, slow growth and declining market shares, particularly in the most advanced and fast-growing industries, go hand in hand with increasing unemployment problems (Table 1.6). There is a growing concern about the direction in which Europe appears to be heading.

Table 1.6 Unemployment as a percentage of the total labour force

	1960–73	1974–79	1980–89	1990–95
Europe (15)	2.3	4.6	9.2	9.8
USA	4.8	6.7	7.2	6.4
Japan	1.3	1.9	2.5	2.5
Korea	na	3.8	3.8	2.4

Source and definitions: OECD Historical Statistics 1960–95.

5 EUROPEAN ECONOMIC INTEGRATION

It is pertinent to ask what European integration has to do with all this. When the first steps towards the present-day European Union were taken, 40–50 years ago, the objectives were partly political and partly economical. On the political front a more stable and peaceful Europe was a primary objective. This obviously remains an important concern for European policy-makers. In the economic area, priority was initially given to the creation of a common market for natural resource-based industries such as the coal and steel industries and agriculture. Because of the political sensitivity of these industries, the steps towards a common market were often combined with subsidies for high-cost, uncompetitive producers, and in many cases these subsidies became of a permanent (or semipermanent) nature. In agriculture, a strong and costly incentive scheme was created, the so-called Common Agricultural Policy (CAP) that encouraged agricultural production. As a result, Europe has become self-sufficient for most agricultural products and a problem of surpluses has occurred, that is being resolved through, among other things, subsidized exports. Hence, it is no surprise that Europe has increased its market share internationally for agriculture and raw materials.

The other main element in the European economic integration process has been a continuous drive towards economies of scale through enlargement and homogenization of markets. After the Second World War various schemes were developed in order to facilitate trade across national borders in Europe. From the late 1950s onwards a process of trade liberalization took place within two European trading blocs, EEC and EFTA, and by the early 1970s internal trade in manufactures was virtually free of tariffs and other restrictions within these two blocs. These developments clearly strengthened the catch-up of Europe in many scale-intensive industries previously dominated by the USA such as cars, domestic electrical equipment and consumer electronics.

The drive towards economies of scale through enlargement of markets continued in the 1970s with the integration of the member countries of the two former trading blocs, EEC and EFTA. By the mid-1980s Western Europe had become a free-trading area for manufactures. However, this failed to produce positive growth effects of the type that had been associated with previous integration efforts. Partly as a result of this the EU launched its plan for a revitalization of the internal European market ('Europe 1992'). This plan was based on the idea that there were large unexploited economies of scale in European industry, the exploitation of which had been prevented by the existence of so-called 'non-tariff barriers' to trade. Hence the heart of the plan was the abolition of these barriers (commonly assumed to be related to discriminatory actions by governments in one way or another). There is evidence that actions to reduce such barriers, in combination with deregulation efforts already under way in many countries, changed the rules of the game in some markets and industries, particularly those that were heavily regulated beforehand (for instance, telecommunications). It is clear, however, that these changes did not have major economic effects of the type foreseen by the European Commission. Perhaps it is still too early to make a final judgement. The fact that Europe's trade performance has been slightly better in scale-intensive industry than in manufacturing as a whole, indicates that although it is difficult to detect an effect on overall growth, the strong emphasis on scale in European integration efforts may have had an impact on its pattern of specialization.

To sum up, European integration has favoured natural resource-based and scale-intensive industries, and this is consistent with the observed change in its pattern of specialization. However, given that modern growth is increasingly knowledge-based, it seems relevant to ask to what extent this move away from the technologically most advanced and fast-growing parts of manufacturing poses a problem for Europe's future growth and welfare.

6 STRUCTURE OF THE BOOK

The remaining chapters focus on Europe's long-term performance along three interrelated dimensions; growth, competitiveness and employment. The first three chapters address these issues from a broad comparative perspective, drawing on experience from Europe, the USA and Japan.

It is often argued that if the prospects for technological progress and productivity growth differ across industries, countries that happen to be specialized in the technologically most progressive industries are likely to get a growth bonus (Lucas, 1988). Conversely, the argument goes, countries that are specialized in the technologically lagging, or 'wrong' industries (Reinert, 1994), tend to do rather badly. Can Europe's performance be explained in this way, or are there other, more generic factors that matter more in the long run? Chapters 2 and 3 both focus on the competitiveness of European industry, but from different angles. While Chapter 2 looks at the record in terms of competitiveness and specialization for major European countries and Europe as a whole, Chapter 3 tries to assess how European industry performs in terms of quality compared to its major competitors in the global market. Among the issues raised are the impact of Europe's pattern of specialization on its growth and trade performance, and the role of R&D, skills and other factors in explaining these developments. Chapter 4 extends this analysis to employment, and asks what the role of technology is for the current unemployment problems in Europe. Is technology a cause – or a cure – for unemployment?

Science-based industry has been singled out as the fastest growing industry in the last decades, as well as the sector in which Europe's performance has been most disappointing. Within the science-based industry the fastest growing has been ICT (information and communication technology). In Chapter 5 we focus on ICT, how radical the changes are and Europe's performance in ICT as a user as well as a producer. What are the causes and consequences of these developments? Is there need for a change of trend, and what can be done about it?

While Chapters 1–5 analyse Europe's long-term performance from a country- or sector-level perspective, Chapters 6–8 focus on the regional level. Chapter 6 looks at European growth at the regional level from the 1950s onwards. It confirms the finding in the literature (Fagerberg and Verspagen, 1996) of very little – if any – convergence in GDP per capita across European regions in the last decades. Various reasons for this finding are explored, including differences across regions in R&D efforts, industrial structure and unemployment. The question is also raised of to what extent there exist groupings of regions in Europe – or 'growth clubs' (Baumol, 1986) – with similar characteristics and dynamics. Is it, for instance, the case – as commonly argued – that the dynamics in southern regions differs significantly from that in the north, or are there other explanations that are equally or more relevant?

A central question in this regard is to what extent differences in regional dynamics in Europe are caused by technological agglomeration, that is, the tendency of technological spillovers to benefit nearby firms rather than firms in more distant locations. The less 'fluid' technology is in this respect, the more likely it is that the economic benefits will concentrate locally. Hence, for policy-making in this area it is of vital importance to come to grips with how technology flows in time and space. Chapter 7 analyses this issue with the help of European patent data (patent citations) broken down on the regional level. Apart from the impact of geographical distance on knowledge flows the chapter also considers other factors such as, for instance, whether knowledge flows more freely within than across national borders.

The tendency towards divergence in regional GDP per capita that can be observed in many European countries in recent years obviously raises a number of questions about policy-making at different levels. Chapter 8 focuses on these questions by means of a detailed comparison of two northern regions from two different countries, both of which are to some extent peripheral. Among the factors discussed are the role of local knowledge infrastructure, particularly universities, in processes of industrial renewal, and the impact of EU regional support schemes. The latter is also addressed on a more general level in Chapter 6.

Chapter 9 again changes the level of analysis by focusing on the challenge for Europe from the dynamics in Eastern Europe and the possible inclusion of these countries into the established system in the West.

Finally, Chapter 10 presents the main conclusions from the study and discusses the policy implications for Europe.

APPENDIX 1.1

The broad product group classification used in Tables 1.3 and 1.4 is based on the 450 product groups of the SIE-World Trade Database, which provides detailed information on export and import of 83 countries. The source for the basic trade statistics of the *SIE-World Trade* is the tapes of the OECD and the UN, according to the three Standard International Trade Classifications (SITC), *Revised, Revision 2, Revision 3*, defined by the Statistical Office of the UN (1961, 1975, 1985 as to the periods 1961–75, 1978–87, 1988 on).

A summary list of the product groups included in each class of products is provided below:

1. *Agricultural products and raw materials* (116 product groups at 4–5 SITC digit level): Food – Live animals – Animal oil and fats – Natural rubber – Meat and meat preparations – Dairy products – Vegetable and fruit preparations – Cereal preparations – Sugar preparations – Other edible products – Vegetable and animal textile fibres – Cork and wood – Skins – Coal – Petroleum oil – Gas – Iron ore – Ores of base metals – Other crude minerals – Paper and paperboard – Petroleum products – Non-metallic mineral manufactures – Non-ferrous metal products.
2. *Traditional industries* (76 product groups at 4–5 SITC digit level): Textile products – Articles of apparel and clothing accessories – Leather manufactures – Footwear – Wood manufactures – Furniture – Paper and printed products – Articles of ceramic materials – Glass products – Miscellaneous manufactures of metal (structures, tools, cutlery and other articles) – Jewellery, goldsmiths – Imitation jewellery – Musical instruments – Sporting goods – Toys & games – Other miscellaneous products.
3. *Scale-intensive* (88 product groups at 4–5 SITC digit level): Organic chemicals – Inorganic chemical products – Other chemical materials and products – Medicinal and pharmaceutical products – Rubber manufactures – Iron and steel – Television, radio, other image/sound recorders and reproducers – Household-type electrical equipment – Ships and boats – Railway vehicles & equipment – Road vehicles.
4. *Specialized suppliers* (43 product groups at 4–5 SITC digit level): Agricultural machinery – Machine tools for working metals – Metal working machinery – Other machine tools for specialized particular industries – Construction and mining machinery – Textile and leather machinery – Paper and paperboard machinery – Other machinery for specialized particular industries – Other general industrial machinery & equipment – Electrical equipment and components – Measuring, checking, analysing instruments – Optical goods – Other miscellaneous products.

5. *Science-based* (59 product groups at 4–5 SITC digit level): Synthetic organic dyestuffs – Radioactive and associated materials – Polymerization and co-polymerization products – Antibiotics and other pharmaceutical products – Nuclear reactors – Automatic data processing machines & units – Telecommunications equipment – Semiconductor devices – Electronic microcircuits – Electronic measuring instruments – Electric power machinery and apparatus – Internal combustion piston engines – Aircraft & associated equipment – Medical instruments – Optical instruments – Photographic apparatus and equipment.

REFERENCES

Abramovitz, M. (1994), 'The Origins of the Postwar Catch-Up and Convergence Boom', in J. Fagerberg, B. Verspagen and N. von Tunzelmann (eds), *The Dynamics of Technology, Trade and Growth*, Aldershot: Edward Elgar, pp. 21–52.

Bank of Korea (1995), *National Accounts 1994*, Seoul.

Baumol, W.J. (1986), 'Productivity Growth, Convergence and Welfare: What the Long Run Data Show', *American Economic Review*, **76**, 1072–85.

Chandler, Alfred D., Jr (1990), *Scale and Scope. The Dynamics of Industrial Capitalism*, Cambridge, MA: The Belknap Press.

Fagerberg, J. and B. Verspagen (1996), 'Heading for Divergence? Regional Growth in Europe Reconsidered', *Journal of Common Market Studies*, **34**, 431–48.

Guerrieri, P. (1992), 'Technological and Trade Competition: the Changing Position of US, Japan and Germany', in M.C. Harris and G.E. Moore (eds), *Linking Trade and Technology Policies*, Washington, DC: National Academy Press.

Johnson, C. (1982), *MITI and the Japanese Miracle: The Growth of Industrial Policy, 1925–1975*, Stanford: Stanford University Press.

Levine, R. and D. Renelt (1992), 'A Sensitivity Analysis of Cross-Country Growth Regressions', *American Economic Review*, **82**, 942–63.

Lucas, R.E. (1988), 'On the Mechanisms of Economic Development', *Journal of Monetary Economics*, **22**, 3–42.

Maddison, A. (1982), *Phases of Capitalist Development*, New York: Oxford University Press.

Maddison, A. (1991), *Dynamic Forces in Capitalist Development*, New York: Oxford University Press.

Mowery, D.C. and N. Rosenberg (1993), 'The U.S. National Innovation System', in R.R. Nelson (ed.), *National Innovation Systems. A Comparative Analysis*, New York and Oxford: Oxford University Press, pp. 29–75.

Nelson, R.R. and G. Wright (1992), 'The Rise and Fall of American Technological Leadership: The Postwar Era in Historical Perspective', *Journal of Economic Literature*, **30**, 1931–64.

OECD (1992), *Technology and the Economy: The Key Relationships*, Paris: OECD.

Pavitt, K. (1984), 'Sectoral Patterns of Technical Change: Towards a Taxonomy and a Theory', *Research Policy*, **13**, 343–74.

Pavitt, K. (1988), 'International Patterns of Technological Accumulation', in N. Hood and J.E. Vahlne (eds), *Strategies in Global Competition*, London: Croom Helm.

Pilat, D. (1993), 'The Economics of Catch-up: the Experience of Japan and Korea', Ph.D. thesis, University of Groningen, The Netherlands, Faculty of Economics, Groningen Growth and Development Centre Monograph Series, no. 2.

Reinert, E. (1994), 'Catching Up From Way Behind. A Third World Perspective on First World History', in J. Fagerberg, B. Verspagen and N. Tunzelmann (eds), *The Dynamics of Technology, Trade and Growth*, Aldershot: Edward Elgar, pp. 168–97.

Rosenberg, N. (1976), *Perspectives on Technology*, Cambridge: Cambridge University Press.

Rosenberg, N. (1982), *Inside the Black Box*, Cambridge: Cambridge University Press.

Van Ark, B. (1993), 'International Comparisons of Output and Productivity', Ph.D. thesis, University of Groningen, The Netherlands, Faculty of Economics, Groningen Growth and Development Centre Monograph Series, no. 1.

von Hippel, E. (1988), *The Sources of Innovation*, New York: Oxford University Press.

von Tunzelmann, G.N. (1995), *Technology and Industrial Progress. The Foundations of Economic Growth*, Aldershot: Edward Elgar.

Wade, R. (1990), *Governing the Market. Economic Theory and the Role of Government in East Asian Industrialization*, Princeton: Princeton University Press.

2. The impact of specialization in Europe

Iain Begg, Bent Dalum, Paolo Guerrieri and Mario Pianta

1 INTRODUCTION

The poor performance of the EU economy over the last two decades, characterized by weak growth and steadily rising unemployment, has prompted research about the competitiveness and specialization patterns of EU industry. This chapter assesses the trade specialization patterns of Europe in the past few decades, a period during which the integration process was consolidated, and examines the impact of specialization on long-run growth.

Europe's trade specialization and performance are analysed with reference, first, to the technological nature of industrial activities and, second, to the sectoral dynamics of world trade. This shows that Europe, overall, has had an unfavourable and sticky composition of trade, but also that there is a great diversity between member countries in their specializations.

Trade patterns matter for growth, as the theoretical and empirical literature has shown. The impact of specialization patterns on European growth is examined in the second part of the chapter, and the link between specialization patterns in trade, production and technology is investigated, together with their consequences for structural change and the economic performances of European countries.

Finally, the policy implications of these findings are discussed, suggesting the need for a greater coherence of innovation and industrial policies, to encourage structural change and the emergence of new high-growth activities in European industries.

2 THE IMPACT OF SPECIALIZATION: A REVIEW OF THE EVIDENCE

The potential for gains from specialization has been central to the development of trade theory since the earliest contributions to the literature and it is an area that has been extensively researched. However, there are many, often conflicting, views about the importance of different forms of specialization and what constitutes a desirable pattern.

Our aim in this chapter is to focus predominantly on specialization in the sense of a relative concentration of output and trade in specific sectors and products. This, in turn, raises the question of what constitutes 'good' or 'bad' specializations. Intuitively, a focus on sectors or products with apparently favourable characteristics would be expected to enhance the economic performance of a country and vice versa.

In much 'standard' trade theory, specialization allows countries to achieve internal economies of scale and greater learning by doing, and thus to increase productivity and grow faster by concentrating on the activities in which they have a comparative advantage. In this approach, productivity growth is derived from specialization 'per se', defined in terms of the degree of dispersion of activities of firms and countries across different fields. In this approach, it does not matter what a country specializes in, because there is always something it can produce and export (Grossman and Helpman, 1991).

An alternative theoretical notion of specialization is based on the recognition that there are systematic differences across sectors and product groups in terms of both productivity and growth potential (Lucas, 1988; Fagerberg, 1996). Specialization in activities characterized by higher rates of productivity growth, higher rates of learning, etc., leads to higher long-run growth. In this approach, the composition of a country's production and its trade specialization are critical. The latter could be of inter-industry type or driven by the increasing variety of activities within each industry, including product differentiation and market segmentation, as well as by the growth of intra-industry trade associated with greater international integration and international production by large multinational firms.

Empirical literature on these different notions of specialization and their impact on growth has offered mixed results. It is useful to summarize briefly some recent empirical contributions in this area.

A group of 39 countries is investigated by Amable (1996), analysing the impact that trade specialization (imports as well as exports) has on productivity growth over the 1965–90 period. Three indicators are used: inter-industry specialization, an index of trade dissimilarity and the role of comparative advantage in electronics. These three indicators account for different dimensions of specialization patterns and combine measures of the degree of a country's concentration of trade across sectors with evidence on the quality of its specialization in the electronics *filière*. All three indicators appear to play a positive role, and a large impact is also found for lagged productivity and the ratio of investment to capital stock, while an education indicator has a negative sign.

Amable's results contrast with those obtained (with a different methodology) by Busson and Villa (1994), from a larger sample of 98 countries. They find a negative impact of inter-industry specialization on growth and their policy conclusions suggest that countries should either specialize in products with fast-growing world demand, or they should increase their intra-industry specialization, which can lead to an increase of overall productivity.

Focusing on OECD countries, the structure and evolution from 1965 to 1992 of inter-industry export specialization patterns are investigated by Dalum, Laursen and Villumsen (1996), addressing both the distribution of countries' exports across industries, and the concentration of OECD exports of individual industries across countries.

The findings point out the stickiness and slow change of specialization patterns, with gradual convergence in the distribution of exports across sectors and countries. This can be attributed to economic convergence among OECD countries, but is somewhat in contrast to the patterns of specialization in technological activities, as measured by patents, which show signs of increasing specialization across sectors (Pianta and Meliciani, 1996).

Patterns of export specialization appear to reflect the specificities of national systems and the persistence of their characteristics in terms of financial systems, education systems, technical infrastructure, R&D systems, etc. (Nelson, 1993; Edquist, 1997; Guerrieri and Tylecote, 1997).

The scope for analysing patterns of specialization is determined by the sectoral breakdown used and it is becoming clear that the available industry classifications are increasingly unable to differentiate the nature of international flows of goods and services. In recent decades, as OECD economies have become more similar, trade within sectors has increased more than traditional inter-industry trade, raising new questions about specialization.

Dalum, Laursen and Verspagen (1996) have addressed the relevance of intra-industry trade specialization with data for 75 product groups assigned to 11 industries. The impact of specialization on growth rates of sectors has been investigated, showing that within-sector specialization in trade has some relevance for growth rates of production, together with variables such as the growth rate of employment, the investment–output ratio, the number of patents per employee, and the gap in productivity (indicating the potential for catching up). However, the ability to explain growth patterns decreases over time and is very limited in the 1980s.

Furthermore, the evolution of high technology industries suggests that while there may be a continuing advantage for the established leaders (who have an absolute as well as comparative advantage), in some cases the 'windows of opportunity' for catching up by (initially) de-specialized countries may be more important than apparent relative advantages of smaller and less active countries. Here again the contrast between the generic specialization, irrespective of what a country is specialized in, and the opportunities trade gives for specialization in certain kinds of activities emerges as an important determinant of growth patterns at the sectoral level. This evidence will be discussed and updated in section 3 below.

To sum up: although many empirical contributions confirm the significance of the relation between specialization and growth, the linkage between the two is still an ambiguous one. Nor has the more recent empirical literature in this field been able to disentangle it.

This evidence points to the limits of generic specialization as a principle for understanding current change, be it in technology, production or trade, and as a policy prescription. While globalization is expected to offer a greater role to countries' comparative advantages per se, the evolution of technology, production and exports appears to be mainly driven by the pattern of structural change, and the specialization in certain activities.

There is some empirical support for the proposition that specialization in specific dynamic sectors aids economic performance, although many contributions also show that the relationship is far from simple. Rather than expecting higher aggregate growth from specialization in whatever field, we have to identify – in the next section – the specific fields where countries show their trade advantages, and consider – in section 4 – the sectoral growth performances resulting from specialization patterns not only in trade, but also in production and technology.

3 EU TRADE SPECIALIZATION TRENDS

3.1 Technology and Specialization of European Countries

This section looks at how the trade specialization patterns of the EU countries have interacted with restructuring and changes in the technological capabilities of the EU in the past few decades as economic integration has been consolidated.

To measure European specialization patterns and their changes over time, we use the *Contribution to Trade Balance* as an indicator. Unlike the more conventional Balassa indicator for Revealed Comparative Advantage (RCA), which is based exclusively on exports, this indicator also takes imports into consideration (see Appendix 2.1).

Structural change and the relationship between technological capability and trade performance in the European economies are analysed using a sectoral taxonomy developed in works by Pavitt (1984) and OECD (1992). This taxonomy is consistent with evolutionary theory on technological change and trade specialization, and identifies five types of industry, primarily through a combination of technology sources, technology user requirements and means of technology appropriation: *natural resource-intensive*, *supplier-dominated or traditional sectors, science-based, scale-intensive* and *specialized suppliers*. All other products are grouped into three main categories: *agricultural products and food industries, other raw materials, fuels*.

The findings for both total EU trade and extra-EU trade (Table 2.1) are that European industry has continued to be specialized in many chemical and mechanical sectors belonging to the *specialized-supplier* and *scale-intensive* groups. In particular EU specialization increased in basic metals and mechanical engineering, such as machine tools and machinery for specialized industries; the latter, it must be recalled, continues to represent vital investment goods for many manufacturing industries (Justman and Teubal, 1991). Foodstuffs and the food industry constitute a special case as they increased their positive contribution to the EC trade balance, as already noted, thanks to a highly protectionist Community Agricultural Policy.

The EU has, by contrast, become less specialized in *traditional* sectors, such as clothing, footwear, light machinery, etc. It has also lost its specialization in *science-based* sectors, and especially in microelectronics and information technologies. Given the prominent role played everywhere by these technologies and products in the restructuring of industries, the relatively weak EU competitive position raises serious concerns. Among these are that:

- *science-based* industries account for large and growing shares of trade and investment in the industrialized area (see Chapter 1, this volume and Fagerberg, 1996);
- they are often the source of important technological innovations, the benefits of which are likely to spill over at intra-industry and inter-industry level (Griliches, 1990; Mohnen, 1990; Scherer, 1982);
- most of them are high-productivity industries and pay higher wages than do other manufacturing sectors (Katz and Summers, 1989).

The relative stability of EU trade patterns contrasts with the dynamic change in Japanese specialization patterns in the same period (Table 2.1). In the early 1970s, Japan's strengths were the *scale-intensive* and *traditional* sectors. Since then, Japan has become increasingly specialized in *special-supplier* and *science-based* sectors, especially in the past decade. The contribution of the scale-intensive sector to the trade balance has been decreasing, and traditional sectors have greatly changed their role.

A comparison between the EU and the USA is also very revealing (Table 2.1). US specialization has long been in R&D-intensive product groups (*science-based*) and this focus has sharpened, confirming the high-level scientific research capability and the propensity of US industry to generate 'primary' innovations. The nature of the relationship between the strengthening of US specialization in R&D-intensive products and the overall US trade

Table 2.1 Patterns of trade specialization – US, Japan and EU*

	United States		
	1970–73	1980–83	1992–95
Agricultural prod.	10.3	10.8	3.9
Fuels	–3.0	–19.6	–6.4
Other raw materials	–1.4	–0.2	0.1
Resource-intensive ind.	–5.8	–4.0	–0.5
Food ind.	–3.2	0.7	2.0
Traditional ind.	–9.6	–5.8	–9.5
Scale-intensive ind.	–6.5	–3.1	–3.9
Specialized supplier ind.	9.8	7.6	3.6
Science-based ind.	11.5	12.9	9.5
Others	–2.2	0.8	1.2

	Japan		
	1970–73	1980–83	1992–95
Agricultural prod.	–24.8	–14.4	–12.5
Fuels	–19.1	–44.9	–16.3
Other raw materials	–11.8	–5.3	–2.9
Resource-intensive ind.	–7.6	–7.5	–5.1
Food ind.	–4.1	–3.4	–7.6
Traditional ind.	9.7	3.9	–9.8
Scale-intensive ind.	46.3	47.7	24.7
Specialized supplier ind.	5.9	11.9	12.4
Science-based ind.	4.9	11.6	17.8
Others	0.5	0.3	–0.6

	EU (15)					
	1970–73	1980–83	1992–95	1970–73 (only extra-EU trade)	1980–83 (only extra-EU trade)	1992–95 (only extra-EU trade)
Agricultural prod.	–6.0	–3.4	–1.9	–14.1	–7.7	–4.8
Fuels	–8.1	–14.3	–4.2	–19.6	–31.3	–10.4
Other raw materials	–1.7	–1.2	–0.6	–3.8	–2.6	–1.4
Resource-intensive ind.	–1.8	–0.8	–0.5	–3.9	–1.9	–0.8
Food ind.	–1.0	1.1	0.8	–2.4	2.4	1.8
Traditional ind.	3.1	2.3	–0.4	6.5	4.5	–2.1
Scale-intensive ind.	9.5	8.7	3.3	22.7	19.5	9.9
Specialized supplier ind.	4.9	5.1	3.6	12.2	11.7	9.1
Science-based ind.	1.4	1.8	–0 3	4.0	4.2	–1.9
Others	–0.3	0.7	0.2	–1.7	1.2	0.6

Notes:
Indicator of comparative advantage (> 0) or disadvantage (< 0). For the formula see text.
* four-year average

Source: UN and OECD trade data from SIE-World Trade Database.

(and industrial) performance is, however, far from clear. There has also been a notable shrinkage of the US trade surplus in the *specialized-supplier* sectors, such as mechanical engineering, although their contribution remained positive up to the mid-1990s.

The EU overall trend, however, is an average that masks sharp differences in the trade patterns of individual EU countries.

Germany's overall competitive position in manufactured products has deteriorated, and this continued into the first half of the 1990s, especially in extra-EU markets. Germany's relatively stable pattern of specialization is broadly based with strengths in both *scale-intensive* industries (especially motor vehicles and transportation equipment, chemicals and related products) and *specialized-supplier* sectors (metalworking machinery, special and general industrial machinery – mechanical engineering generally) (Table 2.2). In comparison with the USA and Japan, Germany has performed relatively poorly in *science-based* products of key importance for 'primary' innovation such as electronics and information technologies.[1]

The *United Kingdom*, on the other hand, had a distinctly negative trade performance, having registered the largest net losses in market shares over the period considered here, which have been spread across a wide range of industries (Guerrieri, 1997). This deterioration in competitiveness is only partially reflected in an unfavourable evolution of specialization pattern in the same period, especially in *scale-intensive* and *specialized-supplier* sectors. Comparative advantage indicators remained strongly positive in science-based sectors, especially in chemical and pharmaceutical industries where the British position was and remained very strong (Table 2.2).

The *French* economy, unlike other European economies, does not have an overarching trade specialization, although there have been significant changes in the period considered here. *Science-based* (high R&D-intensity) sectors have emerged as new strong points of French specialization, with a transfer of resources to these new activities largely made through major public programmes. But they had so far a limited diffusion of technological skill and capabilities outside their sector of origin, as shown by the net losses experienced in most *specialized-supplier* sectors, such as many sophisticated goods belonging to the machinery industries (Table 2.2).

Among the northern European countries, *Sweden* has been able to achieve a significant specialization in a small set of technologically interrelated industries, mostly producing capital and intermediate goods. The list includes a wide variety of *specialized-supplier* industries (in mechanical engineering areas), and many types of *scale-intensive* goods (such as shipbuilding), along with a group of *natural resource-intensive* industries (like pulp and paper). Swedish industry has, however, seen a steady erosion of its share of world exports, which are spread across all the most sophisticated technological sectors

Table 2.2 Patterns of trade specialization – 6 EU countries*

	Germany			France		
	1970–73	1980–83	1992–95	1970–73	1980–83	1992–95
Agricultural prod.	–10.8	–6.8	–3.8	–0.7	1.5	0.9
Fuels	–5.8	–14.7	–4.9	–11.1	–21.6	–5.7
Other raw materials	–2.9	–1.7	–0.8	–1.4	–0.8	–0.4
Resource-intensive ind.	–6.4	–5.9	–2.2	–3.5	–2.0	–2.2
Food ind.	–5.7	–1.6	–1.6	2.8	3.8	2.9
Traditional ind.	–2.9	–2.5	–5.5	3.0	–0.2	–3.7
Scale-intensive ind.	16.9	18.0	10.0	11.3	13.1	3.8
Specialized supplier ind.	14.5	10.8	8.5	–0.2	2.7	–0.1
Science-based ind.	5.3	3.6	0.7	–0.3	3.1	2.7
Others	–2.0	0.8	–0.3	0.0	0.4	1.9
	United Kingdom			**Sweden**		
	1970–73	1980–83	1992–95	1970–73	1980–83	1992–95
Agricultural prod.	–9.9	–4.1	–2.2	–2.8	–3.4	–3.3
Fuels	–8.4	5.5	1.4	–3.2	–13.0	–5.5
Other raw materials	–2.4	–1.6	–0.5	2.5	0.6	0.5
Resource-intensive ind.	–5.0	–2.7	–0.6	5.4	5.3	8.0
Food ind.	–7.8	–3.0	–1.0	–3.7	–2.0	–3.2
Traditional ind.	2.3	–4.2	–4.5	–1.6	–0.1	–3.1
Scale-intensive ind.	14.3	0.8	1.2	2.8	7.9	5.7
Specialized supplier ind.	8.6	4.4	1.9	1.3	4.8	2.8
Science-based ind.	6.8	3.9	3.8	–0.8	–0.4	–2.6
Others	1.5	1.0	0.4	0.0	0.3	0.8
	Italy			**Spain**		
	1970–73	1980–83	1992–95	1970–73	1980–83	1992–95
Agricultural prod.	–10.9	–6.3	–4.6	0.0	–2.5	1.1
Fuels	–14.1	–25.6	–6.0	–12.2	–33.6	–7.8
Other raw materials	–1.5	–1.0	–0.8	–1.9	–1.6	–0.6
Resource-intensive ind.	–0.1	–1.5	–3.5	0.2	4.3	0.9
Food ind.	–5.7	–2.7	–2.9	10.6	5.8	0.7
Traditional ind.	19.1	22.2	19.7	16.4	15.2	2.6
Scale-intensive ind.	6.3	4.5	–4.8	3.5	16.0	9.4
Specialized supplier ind.	6.9	10.0	10.5	–6.8	1.1	–1.4
Science-based ind.	0.8	0.4	–4.3	–8.5	–3.9	–5.3
Others	–0.8	0.1	–3.2	–1.2	–0.7	0.4

Notes:
Indicator of comparative advantage (> 0) or disadvantage (< 0). For the formula see text.
* four-year average

Source: UN and OECD trade data from SIE-World Trade Database.

and which seem to indicate ongoing difficulties in the restructuring and upgrading of trade and technological specialization of the Swedish economy (Table 2.2).

Italy is distinctive in the EU because of its heterogeneous specialization pattern. It has done well in the set of industries where it has traditionally been strong, especially in *traditional* goods ('other industrial products'), and has gained ground in a number of mechanical engineering areas, particularly *specialized-supplier* sectors, like special industrial machinery, and metal-working machinery. This followed extensive restructuring in Italian industry in the past decade, based largely on the adoption and diffusion of (mostly imported) technology, which also allowed some so-called mature sectors to be revitalized. In contrast, Italy is generally losing position in sectors where it has been weak, as in many *science-based* goods (like electronics) and in *scale-intensive* goods (especially chemicals and related products), with negative implications for the future position of Italian industry in world markets (Table 2.2).

Spain, unlike other southern member countries, has seen significant changes in its trade patterns in the period considered here. *Traditional* industries, such as textiles, clothing, footwear and so on, have suffered from the new Asian competition and drastically decreased their contribution to Spain's trade balance. Net exports of foodstuffs, the other strong point of Spanish trade in the 1970s, have also come under pressure in the last ten years. Gains have been made in *scale-intensive* and, to a much lesser extent, *resource-intensive* sectors over the past decade, although there must be doubts about whether these are the industries in which Spain can sustain long-term competitiveness. The data also confirm Spain's lack of presence in *science-based* goods, especially in recent years (Table 2.2).

3.2 Trade Growth and Specialization of European Countries

We turn now to an analysis of the trade specialization patterns of European countries using a different, though complementary, criterion (see Appendix 2.1).

We have already drawn attention to empirical findings which suggest that being specialized in the 'right' kind of activities has become more important for the growth potential of countries over the last decade. The second criterion we investigate here is the growth of world demand, using as a proxy the rate of growth of world exports.

The relation between export growth and specialization at the national level is summarized in Table 2.3. Sustained specialization in high-growth products has mainly been limited to the large OECD economies – the USA, Japan, Germany and the UK, but excluding France. Some smaller economies, notably Ireland, have also been specialized in the 1961–94 high-growth segments, although for Ireland the data show that the transformation happened only after

Table 2.3 EU and major OECD countries' export specialization according to ranking of long-term and medium-term growth rates of exports

	High growth			Medium growth			Low growth		
	1961–73	1973–94	1961–94	1961–73	1973–94	1961–94	1961–73	1973–94	1961–94
USA	0.95	1.16	1.05	1.04	0.75	0.98	1.01	0.92	0.87
Japan	1.44	1.27	1.31	0.95	0.53	0.57	0.56	0.92	0.80
EU	0.98	0.92	0.94	1.03	1.13	1.08	0.98	1.02	1.05
Germany	1.14	1.05	1.07	1.08	0.96	0.88	0.71	0.93	1.02
France	0 90	0.96	0.86	1.06	1.05	1.21	1.01	1.04	1.05
UK	1.03	1.03	1.01	1.03	1.04	1.08	0.93	0.85	0.83
Italy	1.08	0.79	0.95	1.03	1.37	1.00	0.85	1.06	1.18
Sweden	0.98	0.87	0.90	0.79	1.20	1.07	1.36	1.07	1.21
Belgium-Lux	0.98	0.86	0.86	1.00	1.09	1.12	1.02	1.26	1.23
Austria	0.74	0.77	0.86	1.03	1.25	1.09	1.25	1.28	1.31
Denmark	0.85	0.72	0.78	1.13	1.26	1.31	0.98	1.44	1.13
Ireland	0.50	1.20	1.18	1.19	0.75	0.89	1.30	0.78	0.63
Spain	0.80	0.93	0.92	0.91	1.16	1.14	1.38	0.97	0.98
Portugal	0.90	0.61	0.96	0.45	1.64	0 99	1.99	1.17	1.15
Greece	0.28	0.23	0.55	0.74	2.23	1.50	2.25	1.37	1.52

Source: The IKE Trade Database. The ranking of OECD export growth is calculated on exports from all OECD countries. The RCAs are end year values.

1973. This reveals trade patterns broadly similar to the previous analysis of specialization in technology-intensive activities, although Germany, which was not included in the club of countries specialized in high-tech industries, is shown to have specialized in products with rapid growth in exports.

Table 2.4 EU and other major countries' export specialization – three subgroups of the 1961–94 high-growth segment

High-growth subgroups	1961	1973	1984	1994	1961	1973	1984	1994	1961	1973	1984	1994
	USA				*Japan*				*EU*			
(i) Increasing	1.22	1.35	1.81	1.52	0.81	1.36	1.58	1.90	1.00	0.90	0.71	0.72
(ii) Decreasing	0.86	0.79	0.71	0.79	0.91	1.10	1.31	1.18	1.19	1.07	1.00	1.01
(iii) New	1.02	1.07	1.11	1.10	1.05	0.98	1.15	1.16	1.03	1.01	1.00	0.97
ICT	1.11	1.30	1.77	1.48	0.90	1.58	1.79	2.10	1.04	0.89	0.68	0.69
	Germany				*UK*				*France*			
(i) Increasing	1.01	1.01	0.72	0.67	1.17	1.11	1.03	1.26	0.69	0.86	0.73	0.64
(ii) Decreasing	1.72	1.37	1.37	1.26	1.38	0.93	0.63	0.75	1.18	1.07	0.98	0.94
(iii) New	1.19	1.16	1.12	1.07	1.51	1.40	1.16	1.16	0.90	0.93	0.92	0.89
ICT	0.95	0.92	0.64	0.56	1.26	1.09	0.98	1.30	0.75	0.88	0.73	0.62
	Sweden				*Austria*				*Denmark*			
(i) Increasing	1.44	1.43	0.95	0.98	0.42	0.45	0.45	0.44	0.54	0.84	0.57	0.60
(ii) Decreasing	0.85	0.90	1.04	0.93	0.68	0.69	0.78	0.75	0.55	0.60	0.65	0.60
(iii) New	0.58	0.69	0.75	0.81	0.99	1.14	1.19	1.21	0.75	0.87	0.87	1.08
ICT	1.73	1.59	0.96	0.94	0.24	0.34	0.41	0.40	0.32	0.55	0.31	0.48
	Belgium				*Ireland*				*Norway*			
(i) Increasing	0.43	0.51	0.26	0.29	0.05	0.70	3.00	2.23	0.46	0.32	0.20	0.24
(ii) Decreasing	0.63	1.12	1.04	1.26	0.25	0.42	0.58	0.62	0.28	0.25	0.21	0.23
(iii) New	0.70	0.64	0.92	0.71	0.87	1.14	1.01	1.23	0.23	0.35	0.20	0.24
ICT	0.53	0.59	0.27	0.27	0.07	0.68	3.28	2.44	0.42	0.32	0.20	0.23
	Italy				*Spain*				*Portugal*			
(i) Increasing	1.05	0.86	0.53	0.46	0.05	0.25	0.32	0.37	0.06	0.61	0.58	0.20
(ii) Decreasing	1.63	1.27	1.12	1.11	0.19	0.62	0.94	1.35	0.22	0.72	1.13	1.29
(iii) New	1.01	1.25	1.32	1.03	0.34	1.05	0.84	0.74	0.31	0.57	0.71	1.00
ICT	1.31	0.99	0.55	0.46	0.06	0.27	0.37	0.35	0.05	0.77	0.70	0.17

Notes:
The calculations of the RCAs are based on total OECD exports (intra-EU trade included).
The ICT sector is computers, semiconductors and telecommunications.
In the EU aggregate shown intra-EU trade is excluded.
(i) Increasing = High and increasing growth rate of exports.
(ii) Decreasing = High but decreasing growth rate.
(iii) New = New high-growth products.

Source: The IKE Trade Database.

Further insights can be gained by dis-aggregating into the three subgroups of the high-growth segment shown in Table 2.4. These are:

- Products with high and increasing growth rates – that is, high-growth products 1961–94 which increased their ranking by growth rate from 1961–73 to 1973–94;
- Products with high but decreasing growth rates – that is, high-growth products 1961–94 which decreased their ranking by growth rate from 1961–73 to 1973–94;
- Products that became new 'high growers' between 1973 and 1994. Although they have been high-growth products for the entire 1961–94 period, they were not among the high-growth products in the 1961–73 sub-period.

These results facilitate direct comparison of small and large countries. The data show a rather 'bad' pattern for France *vis-à-vis* the other large economies, and the 'good' pattern of Ireland. Among the small countries, the specialization 'scores' for the Netherlands and Sweden are fairly close to 1 in the (i) 'high and increasing group' (and ICT) – a feature which is even more pronounced for Finland; and for all of them clearly higher than the EU average. These countries are heavily specialized in some slow-growing, natural resource-based areas – food in the Netherlands and wood, paper, pulp and related machinery in Sweden and Finland. However, the impact of the presence of a large ICT multinational company in each country (Philips, Ericsson and Nokia, respectively) is visible in the observed patterns.

It might be argued that a specialization in the 'new high growers' group is also desirable, but this depends somewhat on the criteria applied. This is illustrated by Denmark for which the increase in specialization in the new high growers for 1984–94 is accounted for principally by large increases in the share of Danish exports of pharmaceuticals and the rather low-tech-oriented 'other manufactures'.

The same may apply for Austria where the pattern looks pretty bad except for the specialization in 'new high growers'. The latter is based upon an increasing share of pharmaceuticals, power generators and the same residual group mentioned above. What the Danish and Austrian cases underline is that their specialization patterns – reliant hitherto on natural resource-based strongholds (a pattern common to all the small OECD countries) – may be evolving towards segments with higher export growth potential. Their structures may not, therefore, be that 'bad' and they may be moving in the right direction.

Belgium, by contrast, has characteristics which are hard to interpret as 'good'. Its specialization in 'high but decreasing' products has increased but has declined for the 'new high growers'; and the specialization in 'high and increasing' products is very low. More dis-aggregated data show that specialization in 'high but decreasing' is concentrated in organic chemicals, plastics and cars. In all of these Belgium is a hub for assembly by (foreign) multinationals.

When we turn to the European 'NICs', Spain and Portugal both show increasing specialization in the 'high but decreasing' group, but for very different reasons. For Spain, the share of car exports increased from 1984 to the mid-1990s, while the Portuguese RCA of 1.0 in 'new high growers' in 1994 is largely attributable to other manufactures. The inference to draw is that Spain has reached a higher level of development in terms of a more sophisticated export structure, while Portugal relies on less advanced low-tech products, as does Greece.

To sum up, the evidence reported in these two sections reveals the complex character of trade specialization patterns in the EU, and highlights the problems in interpreting them unequivocally. Three trends deserve, however, to be stressed.

The first is the severe loss of specialization and competitiveness experienced in the last decade by the EU countries as a group in *science-based* sectors, and especially in microelectronics and information technology products. In almost every sector of the latter group the trade data clearly demonstrate that European industries are substantially weaker than Japanese and US industries. One of the justifications for the single market programme was, by fostering increased competition, to regain competitiveness in these high-tech sectors. But, on its own, a larger and more competitive European market does not appear sufficient to ensure that European firms achieve higher standards in the electronics and related sectors.

The second point to note is the growing heterogeneity of trade patterns within the EU. Although the integration process has been progressing rapidly in Europe during the past decades, each major European country has a very different sectoral pattern of technological and trade advantage (disadvantage), and national patterns tend to remain stable over time. Both inter-country and inter-industry differences appear to be very important in shaping these national specificities in terms of national competitive advantage and disadvantage.

The third key finding is that there is a persistent and significant north–south difference in specialization patterns within the EU. Richer countries in the north of the Union specialize in technologically relatively advanced sectors, whereas in the southern countries, labour-intensive *traditional* sectors are still the main strengths. These differences are likely to become more significant in the wake of economic and monetary union, since EMU could exacerbate them.

4 SPECIALIZATION IN TECHNOLOGY, PRODUCTION AND TRADE

The detailed evidence in the previous section showed that much of Europe continues to specialize in mature industries, in both the scale-intensive and more traditional sectors, while in high technology EU countries have at best managed to carve out specific niches.

This section extends the analysis in two directions. First, we investigate at the sectoral level whether industries where countries are more specialized show higher than average growth rates. This analysis embraces not only specialization and performance in trade, but also in technology and production, in order to provide a fuller picture of the evolution of the economic activities at the sectoral level (see Appendix 2.1).

In particular, we investigate whether advances in technology and the growth of production and exports have been faster in the sectors where EU countries, as well as the USA and Japan, are strongly specialized. The period of study is 1980 to 1992; the sectoral growth performances of countries are measured by the average annual real growth rates for patents in the USA, value added and exports; specialization indices are calculated for the same three variables.

Two research questions are addressed. The first is whether a relative sectoral strength leads to better performances, as might be expected from traditional theory. In testing this, account is also taken of economic change, which shapes the evolution of overall activities in technology, production and trade.

A second issue explored is the relationships between specialization patterns in technology, production and exports. A relative advantage in technology is expected to be positively associated with better performances in production and exports. The former may sustain the sources of competitiveness resulting in the latter. Equally, a specialization in production may support the national performances in research efforts and exports.

The data for the USA, Japan and the pool of the four major European economies (Germany, France, the UK and Italy) across 19 manufacturing sectors will be analysed. No attempt is made here to develop a comprehensive model explaining sectoral growth patterns; rather, the aim is to provide a systematic picture of the links between specialization and performance in the three indicators described above. Tables 2.5, 2.6 and 2.7 show the regression results for Europe, the USA and Japan, considering as dependent variables the rates of change of value added, exports and patents and investigating the impact

Table 2.5 Specialization and performance in Europe. Four major European economies (Germany, France, UK, Italy), regressions on the pool of 19 manufacturing sectors

Dependent variable	value added rate of change, 1980–82 to 1990–92			exports rate of change, 1980–82 to 1990–92			patents in the USA rate of change, 1980–82 to 1990–92		
	a	b	c	a	b	c	a	b	c
Constant	2.5 ***	1.3 ***	1.2 ***	1.6 ***	0.1	0.5	3.2 ***	2.1 ***	2.1 ***
Spec. index for value added	–5.0 ***	–3.9 ***		–2.2	–0.2		–3.1 **	–0.9	
Spec. index for exports	1.7	0.6		–2.1	–3.6		3.7 **	3.3 **	
Spec. index for patents	0.7	1.7		6.8 **	8.1 ***		–4.3 **	–5.4 ***	
Rate of change of value added					0.2	0.3		0.4 ***	0.4 ***
Rate of change of exports		0.1	0.1					0.1 *	0.04
Rate of change of patents		0.3 ***	0.4 ***		0.3 *	0.1			
Adj. R-squared	0.15	0.26	0.18	0.1	0.12	0.03	0.13	0.27	0.15
F	5.3 ***	6.3 ***	9.0 ***	3.1 **	3.1 **	2.1	4.9 ***	6.6 ***	7.5 ***
Number of cases	76	76	76	76	76	76	76	76	76

Significance levels: *90%, **95%, ***99%

Table 2.6 Specialization and performance in the USA. Regressions on 19 manufacturing sectors

Dependent variable	value added rate of change, 1980–82 to 1990–92			exports rate of change, 1980–82 to 1990–92			patents in the USA rate of change, 1980–82 to 1990–92		
	a	b	c	a	b	c	a	b	c
Constant	3.2 ***	0.44 ***	1.0	3.4 ***	1.7	0.8	4.2 ***	2.9 ***	2.7 ***
Spec. index for value added	–11.3 **	–13.5 **		1.2	3.2		4.0	7.3	
Spec. index for exports	7.6 **	5.9*		4.0	1.6		2.0	–0.8	
Spec. index for patents	0.03	1.2		3.4	4.2		–3.7	–4.0	
Rate of change of value added					0.3	0.2		0.3	0.2
Rate of change of exports		0.2	0.2					0.1	0.2
Rate of change of patents		0.5	0.3		0.2	0.4			
Adj. R-squared	0.15	0.25	0.05	0.04	–0.02	0.09	0.11	0.19	0.08
F	2.1	2.2	1.4	1.0	0.9	1.9	1.8	1.8	1.8
Number of cases	19	19	19	19	19	19	19	19	19

Significance levels: *90%, **95%, ***99%

Table 2.7 Specialization and performance in Japan. Regressions on 19 manufacturing sectors

Dependent variable	value added rate of change, 1980–82 to 1990–92			exports rate of change, 1980–82 to 1990–92			patents in the USA rate of change, 1980–82 to 1990–92		
	a	b	c	a	b	c	a	b	c
Constant	6.0 ***	3.0	–2.9	0.2	–12.3 ***	–4.0	10.3 ***	9.7 ***	7.8 ***
Spec. index for value added	–3.4	0.5		–8.0	–6.2		–0.2	2.6	
Spec. index for exports	3.9	5.0		–4.0	–8.8 ***		3.0	3.9 **	
Spec. index for patents	4.2	–0.7		9.7	6.6		1.1	–2.2	
Rate of change of value added					0.5 *	0.6		0.1	0.4 **
Rate of change of exports		0.5 *	0.2					0.3 **	0.02
Rate of change of patents		0.3	0.8 **		0.9 **	0.1			
Adj. R-squared	0.02	0.42	0.42	0.22	0.66	0.14	0.09	0.49	0.33
F	1.1	3.6 **	7.5 ***	2.8 *	7.9 ***	2.5	1.6	4.5 **	5.4 **
Number of cases	19	19	19	19	19	19	19	19	19

Significance levels: *90%, **95%, ***99%

of the specialization patterns and of other performance indicators. A summary of the results follows:

- *Production* The sectors with the highest growth of value added are those where Europe and the USA have become relatively less specialized, where the fastest technological change is found (as measured by growth in patents) and export performances are strongest (both in terms of rates of growth and in relative advantages). In particular, the changes in value added turn out to be negatively (and significantly) related to the (initial) specialization in production; in Europe and the USA the coefficients are strong and significant both when only the specialization indices are considered in the regression (column a), and when economic and technological changes are included (column b). In Japan the evidence is less clear. Specialization patterns in export and patenting have a positive impact on production performance, but only exports for the US turn out to be significant.

  As expected, growth of value added is positively correlated with indicators of performance in technology and exports. The rates of change of patenting for Europe, and also of exports for Japan, have a strong positive influence on output growth, even when specialization indices are excluded (columns b and c). Economic changes and specialization patterns account for 25 to 40 per cent of total variance.[2]
- *Exports* Perhaps surprisingly, there is a negative correlation for Europe and Japan between export growth and relative export strengths as measured by specialization, though the relationship is only significant for Japan when the economic change variable is included. Technological specialization has a positive influence (significant in Europe), pointing to the importance of technology-based competitive advantages. There is a weak negative relationship between output specialization and export growth, which implies that trends in exports are little affected by the domestic production structures. In the US, on the other hand, all links are positive but not significant. Growth in exports takes place in the sectors where value added and patents also increase, but the evidence of this coherence in the sectoral performances is strongest in the case of Japan alone, where the regression accounts for two-thirds of total variance, while in other cases the explanatory power is much lower.
- *Patenting* There are problems in using patents granted in the USA as the technology indicator for the USA itself, as it has been shown that domestic patenting provides an imperfect picture of sectoral strengths and weaknesses (see Archibugi and Pianta, 1992). It is, therefore unsurprising that the regressions for the USA do not reveal statistically

significant links, although we again find a negative link between technological specialization and patent growth.

In Europe this negative link is stronger and significant, while the evidence for Japan is less clear. Technological performances are always positively affected by export specialization and by the sectoral performances in terms of value added and trade. These variables account for between a quarter and a half of total variance.

Summing up the evidence for the period 1980–92, we find that national sectoral performances in production, export or patenting tend to be negatively correlated with existing specialization patterns. The countries considered, especially the European ones, have grown more in the sectors in which relative specialization has diminished. This suggests *decreasing returns to specialization*, which appears to contradict the promise of increasing benefits from specialization.

The analysis also yields interesting results about the second issue raised – the relationships between specialization patterns in technology, production and trade and overall performances. For Europe and Japan, technological and trade activities appear to have sustained one another, with a secondary role played by production patterns. In the USA these links appear to have been less important (although the caveat about the unreliability of the patenting indicator has to be borne in mind).

The evidence presented here also bears on how specialization can be expected to evolve. For most EU countries, output in fields in which they initially specialized has increased at below-average rates. This should (other things being equal) diminish the specialization and lead to gradual convergence in specialization patterns. However, this tendency may be offset by changes in the global composition of economic activity. If the latter witnesses a growth in sectors in which Europe is relatively under-represented (e.g. electronics), the specialization of Europe in 'traditional', less dynamic industries will increase, with adverse effects on growth. This can be described as a 'vicious circle of specialized decline', as opposed to the 'virtuous circle of dynamic specialization' which occurs when countries concentrate their activities in fast-growth sectors.

The empirical evidence shows some of the limitations of generic specialization as a principle for understanding current economic change and as a policy prescription. While globalization is expected to offer a greater scope for countries to exploit their comparative advantages (measured by their sectoral specializations in technology, production and trade), the negative relationships generally found between specialization and the performance indicators are disturbing. It is also noteworthy that this finding is consistent for all three sets of variables considered, in spite of the differences in the nature of technological accumulation, production activity and export performance.

The conclusions we can draw are that to achieve higher economic growth it is important to be active in the *right*, faster-growing industries rather than to be relatively specialized as such. Large mature economies cannot necessarily rely on trade to yield favourable patterns of specialization. What appears to matter most are the structural shifts brought about by technological and economic change associated with the emergence of new information and communication technologies (ICTs) (see Chapter 5 in this volume).

5 THE IMPLICATIONS FOR EUROPEAN POLICY

Our analysis of EU competitiveness and specialization leads us to draw a number of conclusions. First, the structure and composition of technological activities, of industrial production and trade are important in explaining EU growth and current economic problems, including unemployment (see Chapter 4). Among the key factors (discussed further in other chapters) are the emergence of ICTs and their impact on technological and structural change; the growing role of services; and the decline of the manufacturing base in the EU.

The differences between industries which have been pointed out in this chapter and the apparent decline in European competitiveness in strategically important areas for technological innovation and trade suggest a need for a new generation of industrial and technological policies 'targeting' specific areas with high growth and employment potential.

This is particularly important in ICT-related sectors, where the gap between the growth potential and the lacklustre European performance is greatest. We argue that, when pursuing higher growth, it is not enough to rely on mechanisms such as comparative advantage in increasingly integrated markets. Rather, as we saw in section 2, these are poor predictors of the growth of industries and the process of structural change.

A successful adaptation has to take account of both gains and losses in value added and employment. In general terms, a 'positive' transition to a new pattern of specialization is one in which the net gains exceed what is needed to 'offset' losses, acting through the various mechanisms of redistribution, in order to assure social cohesion.

A critical element will be to pay more attention to innovation in technologies, organizations and institutions. Such a view of the foundations of competitiveness suggests that long-term efforts to adapt the economic structure and to sustain the factors contributing to productivity growth are more relevant than short-term actions aimed at increasing the efficiency of the operation of specific markets.

This is particularly important in the current debate on policies to counter unemployment. Innovation and structural policies provide better long-term solutions than actions focused on the operation of labour markets, such as the attempts to increase flexibility and to assure downward mobility of wages, which have recently been undertaken in most EU countries (see Chapter 4).

Rather, a broader view of innovation might be required, moving from the strict focus on technological innovation to the importance of combining it with organizational innovations in firms and other institutions. To reap all the benefits promised by the diffusion of ICTs and to redistribute them efficiently and effectively across society, a variety of social and institutional innovation might be required (see the final chapter).

APPENDIX 2.1 MEASURING SECTORAL SPECIALIZATION

There are several different measures of specialization. In this chapter we have used two measures.

The first one is the Contribution to Trade Balance (CTB), which is an indicator of trade specialization based on both exports and imports (Section 3.1).

In practice, values greater than 0 (less than 0) of the CTB indicator identify those sectors which give a contribution to the overall trade balance which is higher (lower) than their percentage share in the country's total trade, thus revealing sectors of specialization (de-specialization) in a country (CEPII, 1983; Guerrieri, 1992).

The formula is the following:

$$CTB_j = \left[\frac{(X_{ij} - M_{ij})}{(X_i + M_i)/2} - \frac{(X_i - M_i)}{(X_i + M_i)/2} \times \frac{(X_{ij} + M_{ij})}{(X_i + M_i)} \right] \times 100$$

where X_{ij} = exports of product j of country i; M_{ij} = imports of product j of country i; X_i = total exports of country i and M_i = total imports of country i.

The second measure of specialization used here (sections 3.2 and 4) is the (normalized) revealed comparative advantage (RCA). This is the most commonly used indicator for describing comparative advantages, relative to the world (or other relevant aggregations). The RCA index is obtained by dividing the share of one sector in a country's variable by the share the same sector has in the world (or other aggregate) distribution. It normalizes by the size of countries and of sectors and makes comparisons possible also over time.

In section 3.2 the RCA has been used as an index of trade specialization. The formula is:

$$RCA_{ij} = \frac{X_{ij} / \sum_j X_{ij}}{\sum_i X_{ij} / \sum_i \sum_j X_{ij}}$$

The numerator represents the percentage share of a given sector in national exports – X_{ij} are exports of sector j from country i. The denominator represents the percentage share of a given sector in OECD exports. The RCA index, thus, contains a comparison of national export structure (the numerator) with the OECD export structure (the denominator). When RCA equals 1 for a given sector in a given country, the percentage share of that sector is identical with the OECD average. Where RCA is above 1 the country is said to be specialized in that sector and vice versa where RCA is below 1.

In section 4 specialization has been examined in parallel for technology, production and trade, considering the data for the USA, Japan and the pool of the four major European economies (Germany, France, the UK and Italy) and focusing on manufacturing industry, broken down in 19 sectors. In order to correct for the asymmetric distribution of its values (ranging from 0 to infinity), the specialization index used here is the normalized RCA obtained as (RCA–1)/(RCA+1), and it varies from –1 to +1.

For technology, the number of patents granted in the USA is used; the reference for calculating specialization indices is the total number of patents granted in the USA to inventors of all countries.

For production, value added in constant prices is used, from the OECD STAN database. Due to the lack of sectoral data for total world production, and to lack of constant price data also for several OECD countries, the reference considered for calculating the RCAs has been the total of G6 countries, which accounts for about 80 per cent of OECD value added.

For trade, exports (deflated with the GDP deflator) are considered; the reference for calculating RCAs comes close to the OECD aggregate (includes G6 plus Spain, Denmark, Norway, Sweden and Finland).

Obviously, the lack of a world total may represent a limitation for the calculations of RCAs for value added and exports, but it is unlikely that the results obtained from world data would greatly change the resulting picture.

The period under study is the long cycle of growth from the early 1980s to 1992. Specialization indices have been calculated for both the beginning and the end of the period, but the analysis focuses on the impact the initial specialization profile had on the following period.

NOTES

1. The specialization patterns of each member state in extra-EU trade were also measured, but they do not differ significantly from those shown in Table 2.2.
2. The negative relationship between specialization and performance cannot be explained by the catching-up process, implying that faster growth could be achieved in the sectors lagging more in production activities, or in export and patenting. Inserting a catching-up term in the regressions has never resulted in significant coefficients, except in the case of patenting in Europe where the log of value added per employee turned out to have a positive coefficient, suggesting that technological accumulation and learning by doing were more important than catching-up effects.

   Another extension of the analysis has considered the specialization indices calculated at the end of the period. Japan shows a positive relationship between specialization in production (as well as in exports and patents) and growth, but the coefficients are never significant. Europe and the USA continue to show negative, non-significant links to sectoral growth performances.

REFERENCES

Amable, B. (1996), 'The effects of foreign trade specialisation on growth', paper for the 'Technology, economic integration and social cohesion' project of the EU TSER programme.

Archibugi, D. and M. Pianta (1992), *The technological specialization of advanced countries: A report to the EEC on international science and technology activities*, Preface by Jacques Delors. Norwell, MA and Dordrecht: Kluwer Academic.

Busson F. and P. Villa (1994), *Croissance et specialisation*, Document de travail n. 94, Paris: CEPII.

CEPII (Center for International Prospective Studies) (1983), *Economie mondiale: La montée des tensions*, Paris: Economica.

Dalum, B., K. Laursen and B. Verspagen (1996), 'Does specialization matter for growth?', paper for the 'Technology, economic integration and social cohesion' project of the EU TSER programme.

Dalum, B., K. Laursen and G. Villumsen (1998), 'Structural Change in OECD Export Specialisation Patterns: De-specialisation and "stickiness"', *International Review of Applied Economics*, **12** (3).

Edquist, C (ed.) (1997), *Systems of Innovation: Technologies, Institutions and Organizations*, London and Washington: Pinter.

Fagerberg, J. (1996), 'Technology and Competitiveness', *Oxford Review of Economic Policy*, **12**, 39–5.

Griliches, Z. (1990), 'The Search for R&D Spillovers', *Harvard University Working Papers*, Cambridge, MA: Harvard University Press.

Grossman, G.M. and E. Helpman (1991), *Innovation and Growth in the Global Economy*, Cambridge, MA: MIT Press.

Guerrieri, P. (1992), 'Technological and Trade Competition: The Changing Positions of the United States, Japan, and Germany', in M.C. Harris and G.E. Moore (eds), *Linking trade and technological policies: An international comparison of the industrial nations*, Series on Prospering in a Global Economy, Washington, D.C.: National Academy Press, pp. 29–59.

Guerrieri, P. (1997), 'The changing world trading environment, technological capabilities and the competitiveness of European industry', paper for the 'Technology, economic integration and social cohesion' project of the EU TSER programme.

Guerrieri, P. and A. Tylecote (1997), 'Interindustry Differences in Technical Change and National Patterns of Technological Accumulation', in C. Edquist (ed.), *Systems of Innovation: Technologies, Institutions and Organizations*, London and Washington: Pinter.

Justman, M. and M. Teubal (1991), 'A Structuralist Perspective on the Role of Technology in Economic Growth and Development', *World Development*, Vol. 19, No. 9.

Katz, L.F. and L.H. Summers (1989), 'Industry Rents: Evidence and Implications', *Brookings Papers on Economic Activity: Microeconomics*, pp. 209–75.

Lucas, R. E., Jr (1988), 'On the Mechanics of Economic Development', *Journal of Monetary Economics*, **22**, 3–42.

Mohnen, P. (1990), 'New Technologies and Interindustry Spillovers', *OECD STI Review*, No. 7, pp. 131–47.

Nelson, R. (1993), *National Systems of Innovation: A Comparative Study*, Oxford: Oxford University Press.

OECD (1992), *Technology and the Economy: the Key Relationship*, Paris: OECD.

Pavitt, K. (1984), 'Sectoral Patterns of Technical Change: Toward a Taxonomy and Theory', *Research Policy*, **13**, 343–79.

Pianta, M. and V. Meliciani (1996), 'Technological Specialisation and Economic Performance in OECD Countries', *Technology Analysis and Strategic Management*, **2**, 157–74.

Scherer, F.M. (1982), 'Inter-industry technology flows in the United States', *Research Policy*, **11**, 227–45.

3. European competitiveness: quality rather than price

Marion Jansen and Michael Landesmann

1 TRADE IN GOODS THAT DIFFER IN QUALITY

The traditional view of trade is one of inter-industry trade, with countries exporting goods belonging to one sector and importing goods from another sector. Production costs play a crucial role in explaining this type of trade as countries will export the goods they can produce more cheaply than their trading partners. One would thus expect Spain to export tomatoes to Denmark, because the climate makes it easier and therefore cheaper to produce tomatoes in Spain. In the same way Asian countries export textiles and clothing to Europe because the labour force is cheaper in Asia and thus the cost of producing textiles. This, however, does not seem to be the whole story about trade. A closer look at trade statistics, for instance, reveals that Europe also exports clothing to Asia, among other things expensive designer clothes.

Actually a lot of trade involves the exchange of similar goods. This is in particular the case for trade among industrialized countries. This trade is referred to as being intra-industry trade (IIT) and we distinguish between horizontal and vertical IIT. The first refers to trade in different varieties of the same basic product. Germany's exports of Boss suits and imports of Armani suits fall into that category. Both products are suits, belonging to a certain quality class. They differ mainly in their design. Vertical IIT instead refers to trade in goods of different quality. Asia's exports of cheap no-label clothing for European imports of designer suits falls into the category of trade in vertically differentiated goods. Recent empirical work has suggested that a bigger share of intra-industry trade is vertical as opposed to horizontal (Greenaway, Hine and Milner, 1994; Fontagné and Freudenberg, 1997). It is vertical intra-industry trade, that is, trade in goods of different qualities, that will be the subject of this chapter.

The aim of this chapter is to understand the mechanisms behind vertical intra-industry trade. In particular we will ask the question whether it is the case that countries 'specialize' in the export of certain quality levels. As this turns out to be the case, we will then ask the question what makes a country a high, medium or low quality producer.

Understanding the determinants of product quality can be of much help to policy-makers. For high quality producers it may, for instance, be a bad idea to focus only on the reduction of costs when attempting to gain (retain) market shares. In vertical IIT, product prices must always be seen in relation to product quality. Measures that reduce production costs but at the same time reduce quality, may actually lead to a loss in market share. A country where innovation has created strong quality leaders, should, for instance, be careful when it wants to reduce R&D expenditure in an attempt to lower costs. In a more indirect way changes in labour market regulations could have unforeseen consequences for a country's position in international product markets if these changes reduce the incentives for employers to invest in training of their workers.

Recent theoretical literature has pointed out that market leaders under certain conditions can appropriate higher profit margins than low quality producers (see, for instance, Cabrales and Motta, 1996). In addition, wage claims may be such as to leave producers no option but to locate themselves in the high quality end of the product spectrum if they want to survive. The result is that becoming an exporter of high quality goods would – if successful – improve a country's welfare. Low quality producers may thus wish to understand how they can improve their position and it is therefore important to know which are the factors that affect quality.

2 HOW TO MEASURE PRODUCT QUALITY

Our intention is to analyse the quality of exports at a national level. At this level it is not possible to work with direct measures of product quality such as surveys realized by consumer organizations, as these surveys only exist for a relatively small range of products and it would be difficult to compare them across countries. We will therefore use product prices (values per ton[1]) as an indicator for quality.[2] It seems indeed intuitive to assume that different prices represent different qualities, as consumers would otherwise not want to buy the more expensive good. Yet the use of prices has several drawbacks. First of all it is not obvious which items in the trade statistics should be defined as 'products'. We chose to use a narrow definition of the term 'product' and worked with data at a very high level of disaggregation.[3] But even in this case the 'prices' we calculated may be subject to several distortions. Horizontal differentiation may, for instance, lead to price differences that are not linked to quality differences but simply to cost differences. Additionally, individual product markets may be characterized by different structures of imperfect competition, asymmetric information, consumer loyalty, etc., leading to different price–cost margins. All this may lead to inaccuracies in quality indicators based on price comparisons. The distortions of the aggregate industry and national quality indicators will be negligible as long as the errors in the computed quality indicators at the product level are stochastic; however, it cannot be excluded that the above factors (nature of imperfect competition, etc.) will be systematically related to the specific features of particular markets and market segments and to the characteristics of particular industries.

Throughout this chapter three different 'quality indicators' will be used. These will now be described briefly:

- *the 'price/quality gap' (PG) indicator*[4] The price/quality indicator (*PG*) is a (weighted) average of the ratios at which a particular (national) producer sells its products on EU markets as compared to the prices charged by the mix of its competitors in a particular branch. To construct *PG* for a particular (3-digit NACE) industry *j* we first calculated prices of products *i* at the most detailed product level (in this case 8-digit CN, that is, *Combined Nomenclature*) across the whole range of competitors in the EU markets. A country *c*'s indicator for industry *j* (PG_j^c) was then constructed by taking the ratios of its product prices (p_i^c) relative to the average EU[5] import prices (p_i^{eu}) and aggregating these price ratios for the industry *j*, using the different products' shares in a country's exports (χ_i^c) to the EU as weights. Formally this indicator is derived as:

$$PG_j^c = \sum_{i \in I(j)} \left(\frac{p_i^c}{p_i^{eu}} \right) \cdot \chi_i^c$$

  with *I*(*j*) being the set of items *i* belonging to industry *j*.
- *the 'quality segment' (QS) indicator* The *QS* indicators reveal the biases in the compositions of different country's exports within a particular branch *j* towards high, medium and low quality segments of the product spectrum traded in that branch. To calculate the *QS* indicators we first ranked the prices of individual product items within a particular 3-digit NACE industry (for this ranking we used price data derived for overall trade on EU markets). Then we demarcated three quality segments: Q^I compromising the more highly priced items, Q^{II} the medium and Q^{III} the least priced items. To see whether and in which quality segment national producers specialize, we calculated the ratio of shares of the value of Q^i (i = I,II,III) products in country *c*'s exports to EU markets ($\chi_i^{i,c}$) relative to the share of the value of the Q^i products in total EU imports ($\chi_i^{i,eu}$).[6] The values of *QS* will thus be spread around one and its formal definition is:

$$QS_j^{i,c} = \frac{\chi_j^{i,c}}{\chi_j^{i,eu}}$$

The $\chi_j^{i,c}$ and $\chi_j^{i,eu}$ (i = I,II,III) refer, respectively, to the shares of Q_j^i (i = I high quality, i = II medium quality, i = III low quality) products in country c's exports and in total EU imports within industry j.

- *quality ranks (QR)* For the *QR* indicator we ranked the product prices (this time for 5-digit SITC) that different national producers obtained on the EU import market as a whole. Depending on the exercise, average ranks for a certain industry j ((QR_j^c), with j referring this time to 3-digit SITC and n_j to the number of products in industry j) or for the whole country (QR^c)[7] have been computed. Formally these two indicators are given by:

$$QR_j^c = \frac{\sum_{i \in I(j)} QR^{i,c}}{n_j}, QR^c = \frac{\sum_{j \in J} QR_j^c}{\sum_{j \in J} n_j}$$

When, for instance, the prices of producers from ten countries are ranked, the *QR* indicator would have a value of 5.5 for each country if there was no significant specialization of countries in a certain quality range. The *QR* indicator, however, turns out to differ significantly across countries, with values above 5.5 showing a country's specialization in higher quality segments. If we want to take into account a country's industrial specialization one can again use as weights the shares of different industries/products in a country's exports (the χ_i^c) to arrive at a weighted aggregate quality rank indicator (*WQR*).

The main *difference between the three indicators* is the way in which prices, ranks and/or weights are used. Whereas the *PG* indicator directly uses product prices, the *QR* indicator is based on the ranks of these product prices. To understand the implications of these differences, one has to keep in mind that economic theory predicts that a *higher* price represents a *higher* quality. This means that a price *ranking* should correspond to the quality *ranking* of the underlying products. Which exact price a certain product actually obtains in the market does not, however, only depend on the quality ranking of the product (in the utility space), but also on factors such as differences in the shapes of demand curves for different product qualities; the nature of the market structure in the different quality segments including entry barriers; cross-quality substitution elasticities; and consumer brand loyalties which affect consumer responses to short-term supply price fluctuations due to, for instance, exchange rate movements. Given all these factors which influence price structures across different 'qualities' we experimented also with the *QR* indicators which – by construction – impose a (discrete and uniform) step-wise ranking.[8] The drawback of this indicator is that even small differences in prices will be interpreted as complete steps in a quality ladder, which could also be distorting. When using the *QS* indicator, the focus is on the different commodity compositions of different countries' export baskets. We compare here the shares of a country's exports which fall into the different quality segments. Of course, the choice of three quality segments (rather than more or fewer) is arbitrary. Furthermore, as we compare shares (at current prices) in total export values (in the high, medium and low quality segments), the per unit prices enter (though less directly) as in the *PG* indicator.

Throughout the rest of the chapter the three indicators will be used, though partly for different exercises. Keeping the particular characteristics of each indicator in mind will allow us to develop a rather broad picture of countries' performances in quality competitiveness.

3 QUALITY PERFORMANCES IN EU MARKETS

In the following we shall report the results obtained from applying the above three indicators to detailed EU import data (based on 8-digit CN data in the case of the *PG* and *QS* indicators and 5-digit SITC data in the case of the *QR* indicator).

3.1 Price/Quality Gaps

In Figure 3.1 we can see the development of the price/quality gap indicators for all trading partners in EU markets (that is, EU and non-EU importers) for the group of engineering industries over the period 1988 to 1996. The indicators have been calculated first for each of the 3-digit NACE industries belonging to the engineering group (see note 9) and then aggregated for the group as a whole (using industry shares in a country's exports to the EU as weights). As described above, indicators have been scaled so that they take on the value of 1.0 for total EU imports; values below 1.0 imply that a particular country sells its (weighted) commodity basket on EU markets within that industry at a price below the average of total EU imports (including intra-EU trade); the opposite is true for values above 1.0.

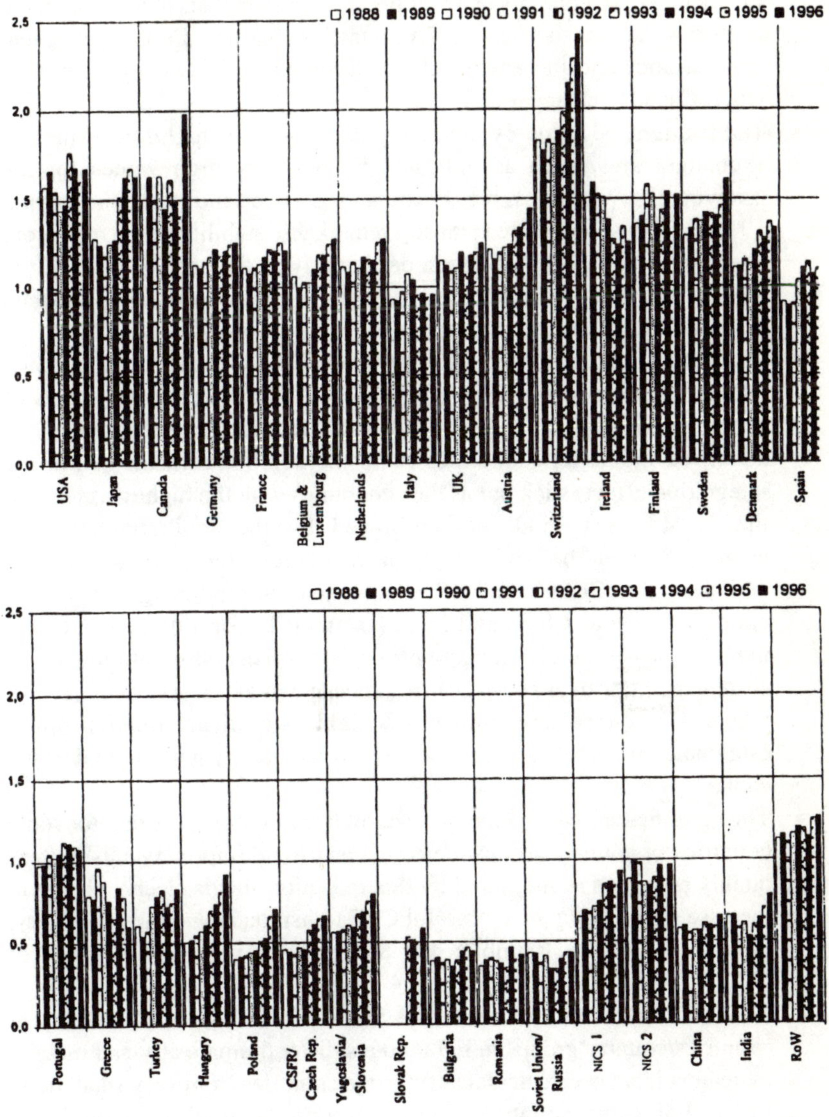

Figure 3.1 Price/quality gap indicators for engineering industries, 1988–96 (EU 12 = 1); exports to EU

The following general results emerge from an inspection of these indicators:

- There seems to be evidence for an EU market integration effect, that is, EU members sell broadly at lower prices on EU markets than comparable countries such as the (ex-)EFTA countries (Austria, Finland, Sweden, Switzerland) and the advanced non-European OECD economies (the USA, Canada and Japan).
- There might be some evidence for the impact of high/low values of exchange rates, such as indicated by the very high values for the price/quality variables for Japan and Switzerland over the period 1993–96. However, there is mostly remarkable stability of the indicators over the years; this can be seen particularly in the case of most of the Central and Eastern European countries (CEECs) which experienced dramatic exchange rate movements over the period.
- From the inspection of the figures we can observe a pattern of 'product quality' ranking in line with prior expectation as regards the stages of economic development associated with the different economies. The advanced industrial economies (with the additional factor of the EU integration effect) stick out as the economies with the highest values for the 'quality gap' indicator, followed by the southern European economies, then the NICs (distinguished here in terms of two groups) and then the CEECs and other low quality producing developing countries such as China and India (see note 10 for a definition of the membership of the different country groups). This observation is given systematic statistical substance in Landesmann and Burgstaller (1998), where – amongst other things – a highly significant relationship is established in panel estimates between price/quality gaps and GDP per capita.
- The other feature which emerges clearly from Figure 3.1 is that for some countries or country groups there are clear trend improvements in their quality positions as measured by this indicator; this is clearly visible in the case of the group of successful CEE transition economies (Hungary, Poland, the Czech Republic and Slovenia) as well as the group of NICs1.
- We should also remind the reader that market shares of the different countries/country groups in EU markets differ dramatically and that the indicators have been calculated for some countries from very small trade flows. For some countries/country groups the market shares have, furthermore, increased dramatically over the period, such as for China and the NICs1 from 0.35 per cent and 0.36 per cent in 1988 to 2.6 per cent and 2.8 per cent respectively in 1996. Such increases have the

> following effects: (1) the calculations for the later years rely on more substantial trade flows, and (2) one can assume that in later periods a broader range of national producers might be represented in those trade flows, while the earlier years might reflect the presence of only a few select exporting firms.

We move on to report results from a first, simple (descriptive) cross-industry regression analysis. Here we used the calculated price/quality gap indicators for the full range of engineering industries (20 NACE 3-digit industries[9]) and the group of textiles, clothing and footwear industries (9 NACE 3-digit industries), then calculated three (or two)-year averages for the periods 1988–90, 1992–94 and 1995–96 and regressed these simply on country (or country group) dummies.[10] These regressions provide an overview of the significance of price/quality gaps across the range of 3-digit NACE industries considered. Comparisons across the periods show, furthermore, interesting shifts in the performances of the different countries or country groups. Tables 3.1 and 3.2 give the results for these and the following regressions reported in this section (for the engineering industries and the textiles, clothing and footwear industries respectively).

Figure 3.2 plots the estimated coefficients on the country dummies for the group of engineering industries for the three periods 1988–90, 1992–94 and 1995–96. We can clearly see the hierarchical range of different quality producers and this range conforms to our expectations. We can see the industrially advanced economies with (mostly) coefficient estimates above one (significance of coefficient estimates above or below one at the 5 per cent significance level have been indicated with an asterisk in Tables 3.1 and 3.2), then moving to coefficient estimates below one for Italy, Spain and Portugal for only the early period, the NICs1 and NICs2 (with coefficient estimates close to one in the last period), to the group of distinctly low quality producers comprising some southern European economies (Greece, Turkey), all of the Central and Eastern European economies, as well as China and India; for this group of low quality producers parameter estimates (below one) were highly significant.

The last three columns in Tables 3.1 and 3.2 present the estimates of whether significant shifts in the estimated country/country group dummies could be detected from one three-year interval to the next three (or two)-year interval. In Table 3.1 we can see that there are significant positive shifts for Japan and many of the 'northern EU' economies as well as Spain (while the position of Greece deteriorates), and substantial improvements by the NICs1 as well as the group of 'western CEECs' and of China, while the quality position of the group of 'eastern CEECs' deteriorates.

Table 3.1 *Coefficient estimates and t-values for the price gap. Results for 3-digit engineering industries*

	Period 1 1988–90	*t*-value	Period 2 1992–94	*t*-value	Period 3 1995–96	*t*-value	Change from period 1 to 2	*t*-value	Change from period 2 to 3	*t*-value	Change from period 1 to 3	*t*-value
USA	1.600	11.02 *	1.593	12.91 *	1.698	10.80 *	–0.007	–0.10	0.106	1.37	0.098	1.16
Japan	1.497	8.03 *	1.781	8.55 *	2.101	6.87 *	0.284	2.57 *	0.320	1.85	0.603	3.95 *
Canada	1.418	5.96 *	1.421	5.51 *	1.551	3.99 *	0.004	0.04	0.130	0.88	0.133	0.94
Germany	1.143	12.24 *	1.205	13.09 *	1.314	8.70 *	0.062	3.20 *	0.109	3.06 *	0.171	5.16 *
France	1.073	3.71 *	1.137	6.39 *	1.232	5.82 *	0.065	2.22 *	0.095	2.26 *	0.159	3.91 *
Belgium/Luxembourg	1.001	0.03	1.107	3.78 *	1.259	6.38 *	0.106	2.58 *	0.152	3.15 *	0.258	5.24 *
Netherlands	1.098	4.34 *	1.180	5.22 *	1.298	6.82 *	0.083	2.01 *	0.117	2.13 *	0.200	4.42 *
Italy	0.904	–4.38 *	0.924	–2.63 *	0.944	–1.93	0.020	0.56	0.020	0.47	0.040	1.12
UK	1.113	4.50 *	1.173	8.02 *	1.255	9.84 *	0.060	1.81	0.083	2.46 *	0.143	3.85 *
Austria	1.242	4.53 *	1.327	4.91 *	1.400	5.63 *	0.085	1.00	0.072	0.73	0.158	1.81
Switzerland	1.898	16.19 *	2.113	16.83 *	2.742	10.33 *	0.215	2.49 *	0.629	3.89 *	0.844	5.42 *
Ireland	1.303	4.48 *	1.258	5.11 *	1.487	5.91 *	–0.045	–0.54	0.230	2.52 *	0.184	1.73
Finland	1.380	4.63 *	1.314	4.57 *	1.470	5.09 *	–0.066	–0.61	0.156	1.38	0.090	0.72
Sweden	1.296	7.64 *	1.412	8.40 *	1.529	7.22 *	0.115	1.84	0.117	1.38	0.232	3.03 *
Denmark	1.087	2.73 *	1.296	4.94 *	1.270	4.32 *	0.210	3.09 *	–0.026	–0.29	0.184	2.84 *
Spain	0.901	–5.81 *	1.037	1.15	1.046	1.26	0.136	3.76 *	0.009	0.19	0.145	3.97 *
Portugal	0.890	–2.20 *	0.955	–1.04	0.927	–1.36	0.065	0.98	–0.028	–0.41	0.037	0.49
Greece	0.831	–3.07 *	0.715	–5.93 *	0.579	–5.94 *	–0.116	–1.59	–0.136	–1.64	–0.252	–2.83 *
Turkey	0.533	–17.40 *	0.564	–13.69 *	0.568	–8.88 *	0.031	0.73	0.004	0.07	0.035	0.66
Hungary	0.441	–23.05 *	0.557	–13.65 *	0.731	–4.16 *	0.116	2.89 *	0.174	2.62 *	0.290	4.80 *
Poland	0.380	–39.51 *	0.454	–27.63 *	0.604	–8.34 *	0.073	2.89 *	0.151	3.27 *	0.224	5.05 *
CSFR/Czech Rep.	0.470	–16.64 *	0.539	–22.90 *	0.721	–5.97 *	0.069	1.83	0.182	3.97 *	0.251	4.60 *
Yu/Slovenia	0.532	–27.52 *	0.603	–17.20 *	0.691	–8.15 *	0.070	2.47 *	0.088	2.11 *	0.159	4.31 *

Slovak Rep.			0.388	–23.52 *	0.438	–11.42 *			0.049	0.97		
Bulgaria	0.432	–9.77 *	0.323	–24.78 *	0.398	–10.53 *	–0.109	–1.70	0.075	1.32	–0.034	–0.40
Romania	0.250	–32.67 *	0.296	–29.10 *	0.311	–20.56 *	0.046	1.37	0.015	0.37	0.061	1.55
SU/Russia	0.547	–5.83 *	0.333	–31.03 *	0.393	–14.28 *	–0.214	–2.64 *	0.060	1.37	–0.155	–1.52
NICs 1	0.738	–14.75 *	0.870	–7.74 *	0.968	–1.13	0.132	5.37 *	0.098	3.17 *	0.230	7.23 *
NICs 2	0.918	–0.89	0.954	–0.69	0.943	–0.96	0.036	0.32	–0.011	–0.12	0.025	0.21
China	0.472	–27.19 *	0.510	–26.58 *	0.559	–20.90 *	0.038	1.44	0.049	1.73	0.087	2.99 *
India	0.686	–5.57 *	0.510	–11.80 *	0.602	–8.84 *	–0.176	–2.53 *	0.092	1.47	–0.084	–1.09
RoW	1.066	1.83	1.077	1.79	1.228	3.53 *	0.011	0.19	0.151	2.02 *	0.161	2.34 *
EU north	1.060	6.14 *	1.146	10.91 *	1.225	13.32 *	0.087	5.23 *	0.078	3.68 *	0.165	9.03 *
EU south	0.875	–4.99 *	0.906	–3.62 *	0.864	–3.83 *	0.031	0.87	–0.043	–1.00	–0.012	–0.27
EFTA	1.455	13.44 *	1.544	14.30 *	1.791	11.37 *	0.089	1.76	0.246	3.34 *	0.336	4.74 *
CEEC 1	0.456	–45.02 *	0.537	–36.83 *	0.686	–12.44 *	0.081	4.63 *	0.149	5.77 *	0.230	9.04 *
CEEC 2	0.419	–16.10 *	0.335	–53.20 *	0.387	–26.08 *	–0.084	–2.43 *	0.052	2.12 *	–0.032	–0.71
NICs	0.828	–3.63 *	0.911	–2.58 *	0.955	–1.36	0.083	1.42	0.044	0.89	0.128	2.02 *

Table 3.2 Coefficient estimates and t-values for the price gap. Results for 3-digit textiles, clothing and footwear industries

	Period 1 1988–90		Period 2 1992–94		Period 3 1995–96		Change from period 1 to 2		Change from period 2 to 3		Change from period 1 to 3	
		t-value		*t*-value		*t*-value		*t*-value		*t*-value		*t*-value
USA	1.281	5.83 *	1.431	7.05 *	1.386	5.13 *	0.151	1.93	–0.045	–0.47	0.105	1.22
Japan	1.855	6.78 *	2.480	6.77 *	2.281	3.71 *	0.624	2.54 *	–0.199	–0.50	0.426	1.29
Canada	1.405	2.80 *	1.225	2.45 *	1.132	1.28	–0.180	–1.05	–0.092	–0.67	–0.272	–1.47
Germany	1.340	14.33 *	1.338	8.87 *	1.403	6.67 *	–0.002	–0.05	0.066	0.95	0.064	1.05
France	1.488	7.68 *	1.511	6.32 *	1.601	7.13 *	0.023	0.23	0.090	0.76	0.113	1.09
Belgium/Luxembourg	1.088	2.71 *	1.178	4.98 *	1.165	4.19 *	0.090	1.86	–0.014	–0.26	0.076	1.50
Netherlands	0.989	–0.40	1.066	1.80	1.132	3.40 *	0.078	1.67	0.066	1.23	0.144	3.05 *
Italy	1.264	6.01 *	1.262	4.44 *	1.334	4.30 *	–0.001	–0.02	0.072	0.75	0.070	0.83
UK	1.258	8.61 *	1.296	9.17 *	1.281	9.87 *	0.038	0.86	–0.015	–0.33	0.023	0.56
Austria	1.570	7.67 *	1.781	6.58 *	1.679	5.39 *	0.210	1.50	–0.102	–0.59	0.108	0.77
Switzerland	2.213	12.46 *	2.333	8.01 *	2.256	5.91 *	0.120	0.62	–0.077	–0.29	0.043	0.20
Ireland	1.106	1.66	1.159	2.21 *	1.068	0.93	0.053	0.56	–0.091	–0.88	–0.037	–0.39
Finland	1.975	5.90 *	1.601	4.19 *	1.409	1.99	–0.373	–1.71	–0.193	–0.79	–0.566	–2.17 *
Sweden	1.322	4.67 *	1.321	3.38 *	1.315	3.94 *	–0.002	–0.01	–0.005	–0.04	–0.007	–0.07
Denmark	1.586	4.65 *	1.496	6.46 *	1.583	6.18 *	–0.089	–0.61	0.086	0.72	–0.003	–0.02
Spain	1.429	3.85 *	1.428	4.11 *	1.487	2.90 *	–0.001	–0.01	0.059	0.31	0.058	0.30
Portugal	1.066	1.95	1.139	1.54	1.085	0.99	0.073	0.76	–0.053	–0.42	0.019	0.22
Greece	1.142	2.05 *	1.211	2.91 *	1.270	3.79 *	0.069	0.69	0.059	0.58	0.128	1.28
Turkey	1.206	1.93	1.108	1.46	0.995	–0.09	–0.098	–0.75	–0.113	–1.17	–0.211	– 1.64
Hungary	0.796	–5.86 *	1.080	1.48	0.954	–1.12	0.284	4.42 *	–0.126	–1.79	0.158	2.94 *
Poland	0.605	–19.38 *	0.782	–6.51 *	0.932	–0.82	0.178	4.54 *	0.150	1.80	0.327	4.20 *
CSFR/Czech Rep.	0.631	–12.34 *	0.819	–5.94 *	0.878	–2.34 *	0.188	4.40 *	0.059	1.02	0.247	4.29 *
Yu/Slovenia	0.903	–2.85 *	1.089	0.96	1.033	0.34	0.186	1.88	–0.056	–0.42	0.130	1.37

Slovak Rep			0.739	–4.29 *	0.810	–2.85 *			0.072	0.79		
Bulgaria	0.455	–13.22 *	0.565	–11.44 *	0.593	–7.39 *	0.110	1.96	0.028	0.43	0.138	2.04 *
Romania	0.621	–10.43 *	0.657	–7.52 *	0.689	–6.53 *	0.036	0.62	0.033	0.49	0.069	1.16
SU/Russia	0.781	–1.17	0.494	–7.78 *	0.536	–6.17 *	–0.287	–1.51	0.042	0.42	–0.245	–1.08
NICs 1	1.034	1.14	1.113	2.27 *	1.034	0.82	0.078	1.36	–0.079	–1.18	0.000	0.00
NICs 2	0.995	–0.09	0.980	–0.26	0.917	–2.16 *	–0.015	–0.16	–0.063	–0.70	–0.078	–1.10
China	0.744	–6.23 *	0.912	–2.52 *	0.910	–3.11 *	0.168	3.10 *	–0.001	–0.03	0.166	3.17 *
India	0.887	–2.24 *	0.870	–3.39 *	0.773	–5.62 *	–0.017	–0.27	–0.097	–1.75	–0.115	–1.72
RoW	0.912	–1.97	0.929	–2.14 *	0.902	–2.54 *	0.016	0.30	–0.027	–0.53	–0.010	–0.17
EU north	1.287	10.90 *	1.307	13.28 *	1.357	12.85 *	0.019	0.56	0.050	1.40	0.070	1.81
EU south	1.212	4.43 *	1.259	4.89 *	1.281	4.05 *	0.047	0.65	0.022	0.25	0.068	0.83
EFTA	1.770	12.21 *	1.759	10.13 *	1.665	7.36 *	–0.011	–0.11	–0.094	–0.81	–0.105	–0.98
CEEC 1	0.734	–13.98 *	0.943	–1.80	0.949	–1.43	0.209	5.62 *	0.007	0.14	0.215	5.59 *
CEEC 2	0.615	–6.03 *	0.610	–14.69 *	0.659	–10.65 *	–0.005	–0.08	0.049	1.18	0.044	0.63
NICs	1.014	0.46	1.046	1.02	0.976	–0.83	0.032	0.58	–0.071	–1.24	–0.039	–0.89

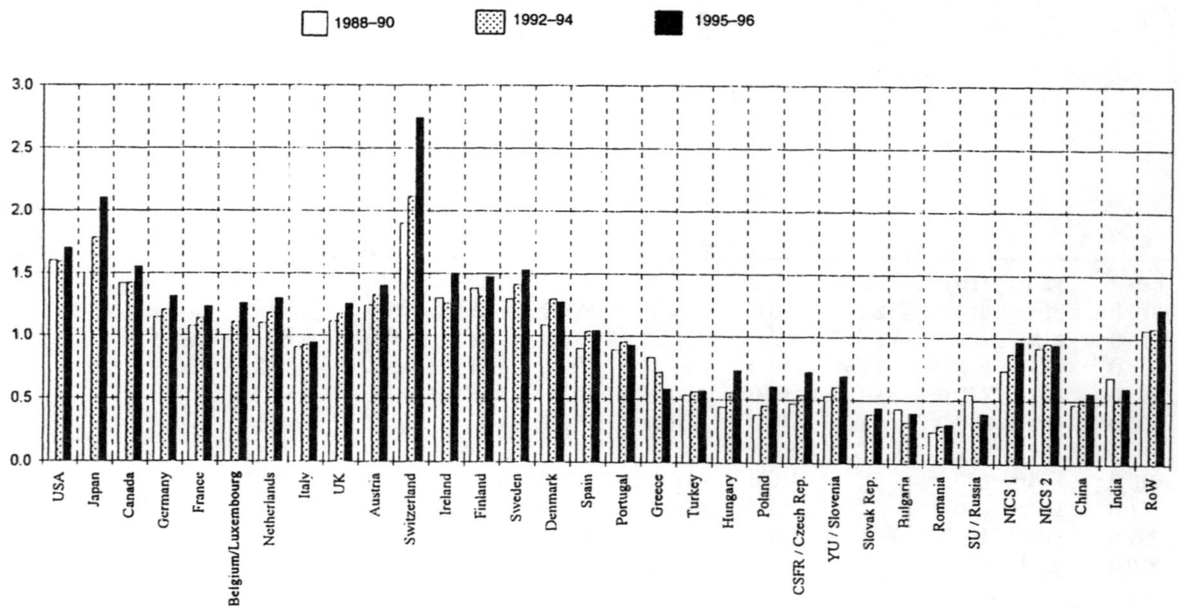

Figure 3.2 *Coefficient estimates for the price gaps. Results for 3-digit engineering industries*

3.2 Representation in the High Product Quality Segments

We now move on to report equivalent results for another quality indicator, *QS*. As discussed previously, the QS^i indicators show the relative representation of different national producers in the high, medium and low quality product segments of a particular industry. Figure 3.3 presents three charts of the periods 1988–90, 1992–94 and 1995–96 for the aggregate of the engineering industries (aggregated again from the quality segmentation analysis conducted at the level of the 3-digit NACE industries). If all three bars for a particular country were of equal size (0.333), then the products belonging to the different quality segments (Q^I for high, Q^{II} for medium, Q^{III} for low quality segments) would be represented in a country's exports to the EU in exactly the same proportions as they are represented in total EU imports (including intra-EU trade). Bars below that value would indicate an 'under-representation' of a country's exports in this quality segment; bars above that value would show an 'over-representation'.

From the charts in Figure 3.3 we can observe the following:

- There is a group of countries for which we observe a consistent (above-average) representation in the highest (Q^I) segment of traded products: this includes the USA, Japan, Canada, Switzerland and Ireland as well as the NICs in the last period.
- The other group which shows a consistent pattern, are the countries which show a consistent (above average) representation of the lowest quality segment (Q^{III}) of traded products in the export baskets to the EU; this group comprises Turkey, all of the Central and Eastern European countries and, over at least two of the three periods, China, India and the NICs2.

To test for the statistical significance of the above observations, we move again to reporting some simple, descriptive regressions which estimate country dummy coefficients in order to find evidence for significant under- or over-representation of a country's exports in particular product quality segments. Tables 3.3 and 3.4 show the estimates of the shares of (high and low) quality products, respectively, in a country's export basket to the EU; we test for significance whether this share is above or below 0.33 (which make up the shares of these products in total EU imports).

From these tables we can see the following:

- Some countries are consistently 'over-represented' in the highest quality segment of trade in engineering products in the EU; these comprise the

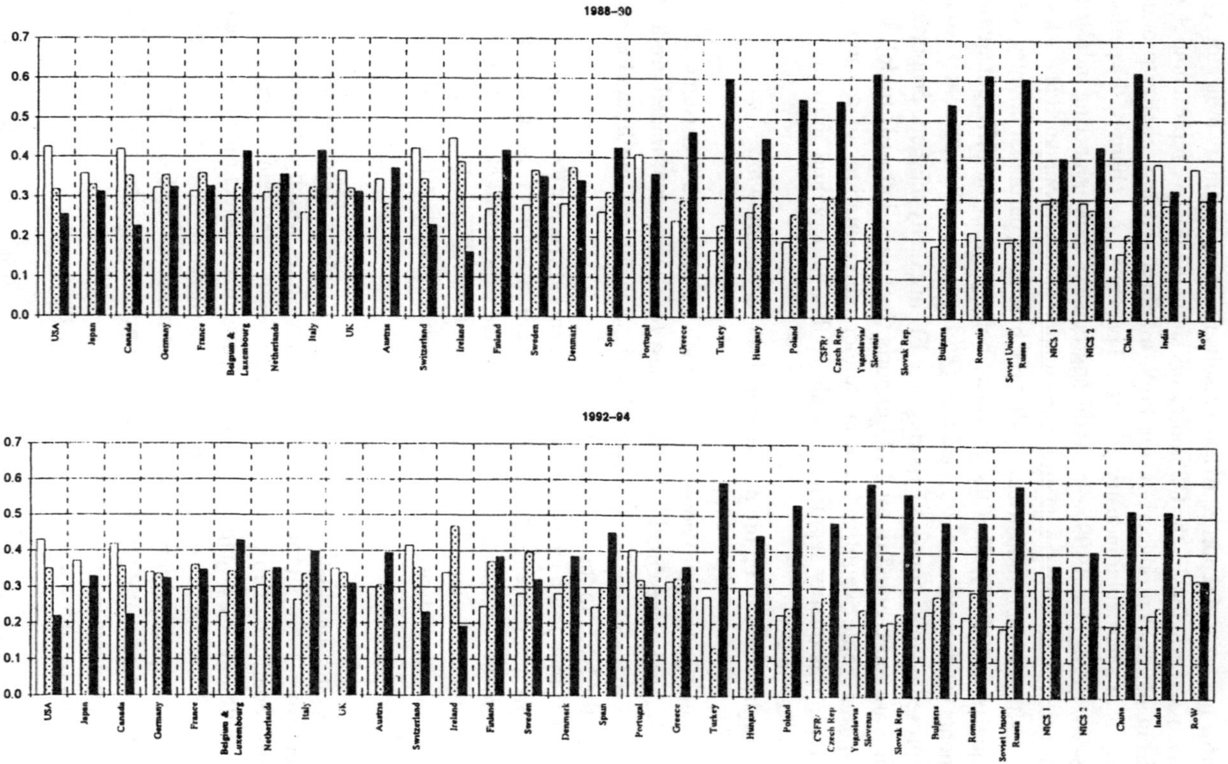
1988–90
0.7
0.6
0.5
0.4
0.3
0.2
0.1
0.0
USA
Japan
Canada
Germany
France
Belgium & Luxembourg
Netherlands
Italy
UK
Austria
Switzerland
Ireland
Finland
Sweden
Denmark
Spain
Portugal
Greece
Turkey
Hungary
Poland
CSFR/ Czech Rep.
Yugoslavia/ Slovenia
Slovak Rep.
Bulgaria
Romania
Soviet Union/ Russia
NICS 1
NICS 2
China
India
RoW
1992–94
0.7
0.6
0.5
0.4
0.3
0.2
0.1
0.0
USA
Japan
Canada
Germany
France
Belgium & Luxembourg
Netherlands
Italy
UK
Austria
Switzerland
Ireland
Finland
Sweden
Denmark
Spain
Portugal
Greece
Turkey
Hungary
Poland
CSFR/ Czech Rep
Yugoslavia/ Slovenia
Slovak Rep
Bulgaria
Romania
Soviet Union/ Russia
NICS 1
NICS 2
China
India
RoW

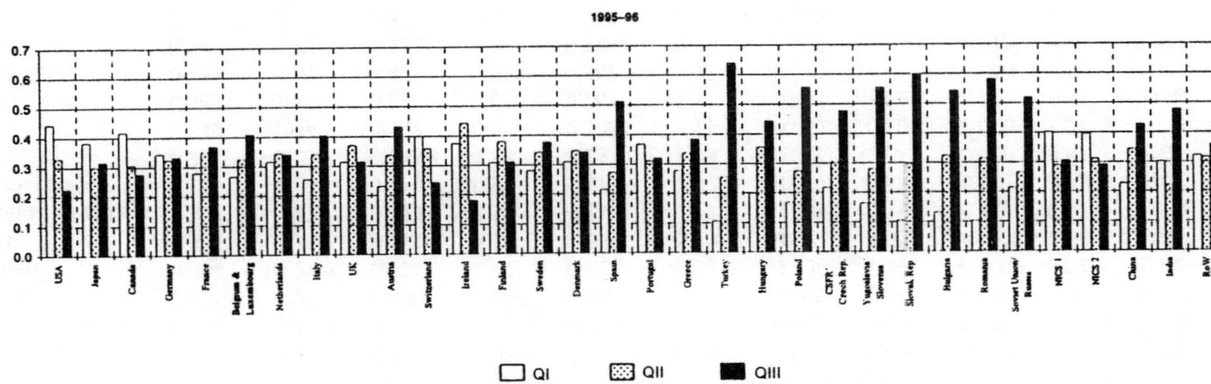

Figure 3.3 *Comparative export structure in different quality segments. Total engineering, 1988–90, 1992–94, 1995–96 1988–90*

Table 3.3 Coefficient estimates and t-values for the high quality segment. Results for 3-digit engineering industries

	Period 1 1988–90		Period 2 1992–94		Period 3 1995–96		Change from period 1 to 2		Change from period 2 to 3		Change from period 1 to 3	
		t-value		*t*-value		*t*-value		*t*-value		*t*-value		*t*-value
USA	0.431	5.94 *	0.411	4.93 *	0.436	6.59 *	–0.020	–0.90	0.025	1.10	0.005	0.20
Japan	0.387	2.55 *	0.393	2.76 *	0.470	4.76 *	0.007	0.22	0.077	2.18 *	0.084	2.41 *
Canada	0.428	3.99 *	0.430	3.89 *	0.361	1.09	0.002	0.06	–0.070	–1.91	–0.068	–1.91
Germany	0.304	–3.24 *	0.324	–0.77	0.321	–0.98	0.020	1.39	–0.004	–0.22	0.016	1.09
France	0.305	–2.56 *	0.281	–5.26 *	0.265	–5.70 *	–0.023	–1.55	–0.017	–1.06	–0.040	–2.37 *
Belgium/Luxembourg	0.278	–3.47 *	0.274	–3.50 *	0.286	–2.17 *	–0.004	–0.18	0.013	0.47	0.009	0.33
Netherlands	0.333	–0.03	0.316	–0.97	0.316	–1.07	–0.017	–0.67	0.001	0.03	–0.017	–0.65
Italy	0.261	–5.23 *	0.256	–4.88 *	0.235	–6.03 *	–0.005	–0.24	–0.020	–0.87	–0.025	–1.18
UK	0.340	0.64	0.328	–0.50	0.308	–2.11 *	–0.012	–0.80	–0.020	–1.27	–0.032	–2.01 *
Austria	0.363	1.41	0.312	–1.03	0.263	–3.51 *	–0.051	–1.72	–0.049	–1.64	–0.100	–3.32 *
Switzerland	0.433	4.43 *	0.401	2.92 *	0.361	0.95	–0.032	–1.00	–0.040	–1.10	–0.072	–2.01 *
Ireland	0.392	2.77 *	0.362	1.23	0.346	0.40	–0.029	–0.93	–0.016	–0.41	–0.045	–1.23
Finland	0.315	–0.70	0.270	–2.83 *	0.330	–0.10	–0.045	–1.32	0.059	1.53	0.015	0.35
Sweden	0.297	–2.04 *	0.265	–3.84 *	0.283	–2.10 *	–0.032	–1.29	0.018	0.60	–0.015	–0.50
Denmark	0.290	–2.22 *	0.260	–3.48 *	0.284	–2.33 *	–0.030	–1.04	0.024	0.78	–0.006	–0.20
Spain	0.277	–3.06 *	0.258	–4.93 *	0.213	–7.04 *	–0.019	–0.81	–0.044	–1.91	–0.064	–2.43 *
Portugal	0.314	–0.72	0.325	–0.33	0.298	–0.87	0.010	0.28	–0.027	–0.57	–0.016	–0.35
Greece	0.285	–1.66	0.351	0.55	0 339	0.14	0.066	1.53	–0.012	–0.23	0.054	1.11
Turkey	0.217	–4.63 *	0.251	–2.74 *	0.207	–4.39 *	0.034	0.87	–0.044	–1.01	–0.010	–0.26
Hungary	0.261	–2.67 *	0.261	–2.63 *	0.191	–6.75 *	0.000	–0.01	–0.070	–1.86	–0.070	–1.90
Poland	0.192	–6.66 *	0.196	–7.93 *	0.155	–9.98 *	0.004	0.14	–0.040	–1.58	–0.036	–1.24
CSFR/Czech Rep.	0.167	–8.95 *	0.233	–4.77 *	0.276	–1.60	0.066	2.35 *	0.042	1.08	0.108	2.89 *
Yu/Slovenia	0.163	–9.93 *	0.174	–8.97 *	0.190	–3.88 *	0.011	0.46	0.016	0.43	0.027	0.74

Slovak Rep.			0.207	–6.33 *	0.173	–5.01 *			–0.035	0.97		
Bulgaria	0.236	–3.14 *	0.275	–1.87	0.180	–4.29 *	0.039	0.90	–0.096	–2.01 *	–0.056	–1.19
Romania	0.194	–3.94 *	0.205	–4.25 *	0.211	–2.93 *	0.011	0.24	0.006	0.11	0.017	0.30
SU/Russia	0.238	4.13 *	0.247	–3.74 *	0.301	–1.00	0.009	0.27	0.054	1.38	0.063	1.61
NICs 1	0.325	–0.41	0.301	–1.68	0.323	–0.42	–0.024	–0.87	0.022	0.74	–0.002	–0.05
NICs 2	0.361	0.82	0.340	0.22	0.335	0.04	–0.020	–0.45	–0.005	–0.12	–0.026	–0.53
China	0.246	–3.43 *	0.214	–5.73 *	0.232	–3.56 *	–0.032	–0.96	0.019	0.54	–0.013	–0.34
India	0.330	–0.15	0.260	–2.93 *	0.277	–1.89	–0.070	–1.94	0.017	0.45	–0.052	–1.33
RoW	0.381	2.72 *	0.365	1.57	0.360	1.12	–0.015	–0.58	–0.005	–0.15	–0.020	–0.70
EU north	0.301	–5.69 *	0.291	–7.08 *	0.288	–7.13 *	–0.010	–1.25	–0.003	–0.38	–0.014	–1.58
EU south	0.292	–2.85 *	0.311	–1.50	0.283	–2.44 *	0.019	0.91	–0.027	–1.11	–0.009	–0.35
EFTA	0.352	1.65	0.312	–1.93	0.309	–1.77	–0.040	–2.52 *	–0.003	–0.18	–0.043	–2.42 *
CEEC 1	0.196	–12.61 *	0.216	–10.85 *	0.203	–8.73 *	0.020	1.32	–0.013	–0.72	0.007	0.41
CEEC 2	0.223	–6.35 *	0.234	–7.51 *	0.216	–6.46 *	0.011	0.51	–0.018	–0.81	–0.007	–0.27
NICs	0.343	0.49	0.321	–0.69	0.329	–0.21	–0.022	–0.83	0.008	0.30	–0.014	–0.48

Table 3.4 Coefficient estimates and t-values for the low quality segment. Results for 3-digit engineering industries

	Period 1 1988–90		Period 2 1992–94		Period 3 1995–96		Change from period 1 to 2		Change from period 2 to 3		Change from period 1 to 3	
		t-value		*t*-value		*t*-value		*t*-value		*t*-value		*t*-value
USA	0.258	–5.01 *	0.252	–5.40 *	0.248	–4.52 *	–0.006	–0.27	–0.004	–0.17	–0.010	–0.41
Japan	0.259	–4.10 *	0.264	–3.89 *	0.225	–5.06 *	0.005	0.18	–0.039	–1.40	–0.035	–1.23
Canada	0.251	–3.90 *	0.270	–2.61 *	0.297	–1.34	0.019	0.57	0.027	0.74	0.046	1.35
Germany	0.350	1.29	0.346	0.85	0.352	1.00	–0.004	–0.21	0.006	0.27	0.002	0.10
France	0.355	1.73	0.379	3.31 *	0.402	4.25 *	0.024	1.31	0.023	1.08	0.047	2.36 *
Belgium/Luxembourg	0.400	3.80 *	0.409	4.14 *	0.372	1.80	0.009	0.35	–0.036	–1.28	–0.028	–1.00
Netherlands	0.354	1.21	0.358	1.64	0.347	0.81	0.004	0.18	–0.011	–0.47	–0.007	–0.27
Italy	0.419	5.05 *	0.426	4.63 *	0.462	6.28 *	0.007	0.25	0.036	1.24	0.043	1.62
UK	0.320	–1.22	0.322	–0.98	0.359	1.36	0.002	0.15	0.037	1.78	0.039	1.91
Austria	0.360	1.33	0.381	2.64 *	0.412	3.00 *	0.021	0.79	0.031	1.01	0.052	1.62
Switzerland	0.233	–5.41 *	0.256	–3.88 *	0.286	–1.73	0.023	0.85	0.030	0.91	0.053	1.68
Ireland	0.278	–2.58 *	0.286	–1.75	0.314	–0.59	0.008	0.23	0.028	0.67	0.036	0.96
Finland	0.380	1.87	0.384	2.19 *	0.363	1.13	0.004	0.11	–0.021	–0.59	–0.017	–0.47
Sweden	0.378	1.99	0.365	1.46	0.404	2.87 *	–0.013	–0.42	0.040	1.21	0.027	0.79
Denmark	0.378	2.01 *	0.408	3.02 *	0.373	1.56	0.030	0.90	–0.035	–0.97	–0.005	–0.15
Spain	0.416	3.79 *	0.419	4.56 *	0.492	7.06 *	0.003	0.10	0.073	2.50 *	0.076	2.36 *
Portugal	0.454	4.41 *	0.414	2.99 *	0.421	2.07 *	–0.040	–1.05	0.007	0.14	–0.034	–0.70
Greece	0.383	1.73	0.317	–0.63	0.386	1.27	–0.066	–1.70	0.070	1.48	0.004	0.08
Turkey	0.510	4.95 *	0.501	4.84 *	0.561	6.42 *	–0.009	–0.19	0.060	1.17	0.050	0.97
Hungary	0.451	3.56 *	0.451	3.84 *	0.508	5.20 *	0.000	–0.01	0.057	1.24	0.057	1.16
Poland	0.539	6.59 *	0.548	8.68 *	0.571	7.50 *	0.009	0.23	0.023	0.58	0.032	0.70
CSFR/Czech Rep.	0.541	6.17 *	0.495	5.63 *	0.471	4.41 *	–0.046	–1.04	–0.024	–0.56	–0.070	–1.46
Yu/Slovenia	0.576	9.30 *	0.565	7.25 *	0.506	3.89 *	–0.011	–0.27	–0.059	–1.10	–0.070	–1.44

Slovak Rep.	.		0.496	5.37 *	0.585	5.69 *			0.089	1.73		
Bulgaria	0.439	2.98 *	0.466	3.71 *	0.510	4.16 *	0.027	0.53	0.044	0.78	0.070	1.27
Romania	0.584	5.90 *	0.503	4.50 *	0.533	4.27 *	–0.081	–1.42	0.030	0.50	–0.051	–0.79
SU/Russia	0.448	3.11 *	0.490	4.86 *	0.430	2.48 *	0.042	0.87	–0.060	–1.19	–0.018	–0.32
NICs 1	0.374	1.67	0.365	1.38	0.317	–0.72	–0.009	–0.27	–0.048	–1.45	–0.057	–1.64
NICs 2	0.333	–0.02	0.348	0.47	0.326	–0.22	0.015	0.35	–0.022	–0.47	–0.007	–0.15
China	0.464	3.82 *	0.476	4.77 *	0.421	2.58 *	0.011	0.25	–0.054	–1.19	–0.043	–0 86
India	0.332	–0.04	0.448	3.27 *	0.461	2.89 *	0.116	2.52 *	0.012	0.22	0.129	2.50 *
RoW	0.323	–0.55	0.339	0.29	0.355	0.98	0.015	0.60	0.016	0.57	0.031	1.11
EU north	0.368	5.56 *	0.378	6.68 *	0.381	6.14 *	0.010	1.12	0.003	0.29	0.013	1.33
EU south	0.418	5.57 *	0.383	3.47 *	0.433	4.67 *	–0.034	–1.65	0.050	2.01 *	0.015	0.60
EFTA	0.338	0.39	0.346	1.22	0.366	2.45 *	0.009	0.55	0.020	1.16	0.029	1.62
CEEC 1	0.527	12.27 *	0.515	12.28 *	0.514	10.11 *	–0.012	–0.56	–0.001	–0.03	–0.013	–0.53
CEEC 2	0.490	6.95 *	0.489	9.15 *	0.515	8.29 *	–0.002	–0.06	0.026	0.94	0.024	0.77
NICs	0.353	1.00	0.356	1.19	0.321	–0.59	0.003	0.12	–0.035	–1.22	–0.032	–1.09

USA and Japan, as well as Canada and Switzerland over the first two periods.

- There are a number of EU economies which show a significant under-representation in the highest quality segment of engineering products trade in EU markets and these comprise France, Belgium/Luxembourg and Italy, as well as Spain and Denmark.
- Much larger gaps can again be found for Turkey and the countries of Central and Eastern Europe as well as China.
- A look at Table 3.4 confirms a more or less symmetric picture with regard to the over-/under-representation in the lowest quality segment of engineering products traded in EU markets. Here the USA, Japan and Switzerland show a significant 'under-representation'; France, Italy and Austria a significant over-representation; and this over-representation is accentuated in the case of Spain and Portugal and more so in the case of Turkey and the countries of Central and Eastern Europe.
- As to estimates of shifts in the estimated parameters of the country dummies over the defined sub-periods, we find much fewer significant shifts here than we found with the other 'price/quality' gap indicator (*PG*); the reason for this could be that a shift here means changes in the composition of the product basket exported to EU markets while a shift in the price/quality gap indicator means simply a change in the price index of a given basket of commodities; the former phenomenon might be a longer-term structural adjustment in the product programme supplied by producers than the latter.

3.3 Quality Ranks

Table 3.5 presents the quality ranks of the different national exporters to EU markets using the *QR* and the weighted *QR* indicators (*QRW*) for the years 1988 and 1992 respectively. Amongst the EU economies, Germany achieves the highest quality ranking, followed by Ireland which in its export sector is strongly dominated by transnational corporations. There is again evidence of high quality ranking by some of the then EFTA economies (Switzerland, Sweden, Austria) which achieve a quality ranking somewhat in line with the USA and Canada. Japan's rank is in line with the bulk of the EU economies and Singapore's position somewhat above it; the rest of Asia is clearly below these two economies with a clear gradation across these (much less visible amongst the EU economies). The countries of Central and Eastern Europe achieve – in accordance with the other two quality indicators – rankings which are amongst the lowest of all exporters to EU markets. The lower positions of the Asian economies as compared to the measures obtained with the other quality indicators are mostly due to the fact that we are here compiling an indicator for the economy as a whole, while the other indicators were calculated for sub-sectors (engineering, and textiles, clothing and footwear).

Table 3.5 Quality ranking of national producers in EU markets

	QR88	QR92W	QR92
France	20.65	21.24	21.32
Bel/Lux	19.82	21.08	21.28
Netherlands	19.29	18.58	19.82
Germany	22.77	23.91	24.12
Italy	19.62	21.17	21.03
UK	21.26	20.67	21.08
Ireland	23.93	22.63	21.86
Denmark	19.99	20.45	20.58
Portugal	19.00	20.06	20.52
Spain	16.70	18.20	18.10
Norway	20.37	20.53	20.93
Sweden	22.55	23.37	23.82
Finland	18.45	17.46	18.42
Switzerland	26.29	26.90	26.80
Austria	22.74	22.87	22.92
Turkey	12.89	14.72	15.19
USA	24.44	23.04	23.08
Canada	20.62	20.35	22.10
Australia	22.50	18.35	20.26
New Zealand	15.06	8.90	14.89
China	10.30	10.48	10.41
South Korea	12.86	14.81	15.73
Japan	20.59	20.77	20.53
Taiwan	13.04	14.07	13.41
Hong Kong	13.62	13.48	14.99
India	12.83	11.29	11.35
Thailand	13.20	13.13	13.78
Indonesia	12.17	12.18	9.86
Malaysia	14.14	15.07	15.07
Singapore	18.65	20.74	22.32
Poland	11.39	12.07	12.94
Czech R.	10.11	10.18	11.19
Hungary	14.40	15.24	16.46
Romania	10.02	8.84	10.23
Bulgaria	11.13	9.25	9.76
Sov. Union	12.94	11.06	10.18

3.4 Summarizing the Preliminary Evidence on the Quality Ranking in EU Markets

We summarize the most important findings of our preliminary statistical analysis so far:

- The evidence suggests a clearly marked pattern of vertical product differentiation in EU product markets insofar as this could be revealed with the partial data set on trade flows (including intra-EU trade) in the EU markets.
- There is some casual evidence that the ranking of a country's position in these vertically differentiated product markets is linked to its stage of economic development in general (as measured by GDP per capita).
- There is evidence for some significant differentiation both across EU-north and EU-south and between the advanced non-EU and EU economies. In the engineering branches, the USA and Japan reveal a significant positive price gap in relation to total EU imports (and so do the EFTA countries, particularly Switzerland) and a significant under-representation in the low quality segment; the EU-south, on the other hand, reveals a significant price/quality discount on its exports (to EU markets) and – together with the EU-northern countries, for the latter basically due to Belgium and Italian export structures – a significant under-representation of the high quality segment in its overall exports. In the textiles, clothing and leather branches, on the other hand, no such differentiation between the advanced EU and non-EU economies could be detected; both these groups of economies sell at significantly higher unit prices than their lower quality competitors in this area (and so do the EFTA countries). However, the overall explanatory power of our econometric estimates for the quality segmentation analysis in these branches is significantly lower than for the engineering branches.
- Finally, there is evidence that the increased presence of Central and Eastern European producers as suppliers of engineering products to EU markets has significantly extended the range of countries situated at the low quality end of EU engineering imports.

3.5 Differences Amongst EU and Non-EU Advanced Economies

What do we make of the differences between the EU and non-EU (particularly the USA and Japan) advanced economies? Here we would like to insert a note of caution in interpreting these results as revealing significant quality deficits between EU and non-EU producers. We should reiterate that we are looking at the competition of producers in EU markets: one would expect that EU

Table 3.6 Comparison of price gaps US–EU, 1995/96; (a) engineering industries in bilateral trade flows; (b) in total EU imports

	USA	Germany	France	Belgium and Luxembourg	Netherlands	Italy	UK	Ireland	Denmark	Spain	Portugal	Greece
(a) 1995	1.023	1.363	1.275	1.303	1.342	0.849	1.113	0.996	1.322	0.555	0.707	0.864
1996	1.035	1.534	1.328	1.368	1.303	0.850	0.979	1.551	1.303	1.017	0.834	0.779
(b) 1995	1.607	1.248	1.255	1.249	1.255	0.938	1.199	1.342	1.371	1.105	1.096	0.781
1996	1.678	1.257	1.210	1.275	1.276	0.959	1.250	1.255	1.338	1.096	1.083	0.603

producers would have preferential access to these markets, given the degree of product market integration within the EU, attempts to harmonize production standards, removal of many types of entry barriers across EU firms, etc. This would mean that a segment of less competitive firms (lower quality suppliers) would still be able to sell their products on EU markets, which would not be the case for non-EU suppliers. The average performance of EU economies in such a comparison would thus show that a wider range of EU firms (including the segment of lower quality suppliers) would be represented in EU markets than would be the case for non-EU exporters. Expressed differently, on EU markets there would be a bias for non-EU economies to be represented (relatively) by the top end in the distribution of firms, while EU producers would be represented by a wider range in the distribution of firms. The rather high values obtained for EFTA countries (which are in line with the estimates obtained for the USA and Japan) seem to support this view.

To test this hypothesis further, we undertook another exercise and calculated for the years 1995 and 1996 another set of *PG* indicators from trade flows between EU economies and the USA. Hence, we pooled the bilateral trade data between each EU country and the USA. This allowed us to compare the performances of the set of EU firms selling to the US market with the set of US firms selling in EU markets, and hence amounted to correcting for the 'EU home market advantage' of EU suppliers. In Table 3.6 we present in a comparative way the *PG* indicators obtained for the aggregate of the engineering industries (constructed again from detailed estimates at the 3-digit NACE level) from both types of exercises: the relative performances of the US and EU economies in EU markets (our earlier exercise) and the relative performances of the US and EU economies in the bilateral trade flows between the US and the EU economies. We can see that the performance of the US relative to the EU economies drops dramatically in the latter exercise and is often below that of most of the northern EU economies. This supports our hypothesis that the relatively bad performance of EU economies in EU markets observed in our earlier exercise is most likely due to the selectivity effect which lower barriers of entry exert on the range of EU firms represented in intra-EU trade as compared to the non-EU companies represented in EU markets.

4 THE DETERMINANTS OF PRODUCT QUALITY

The previous sections have shown that countries do indeed tend to specialize in certain quality segments. The aim of this section is to find out which are the factors that make a country a high or a low quality producer. An answer to this question will help policy-makers understand how certain measures may affect the international competitiveness of their firms. In particular it indicates policy directions to those countries that wish to raise the quality level of their exports.

When a firm decides whether to produce a high or low quality product it will in the first place take into account the costs of producing a certain quality level. In countries where factors necessary for the 'production of quality' are readily available, firms will tend to be more willing to invest in the production of high quality goods. This is the case because the large supply of these factors makes them relatively cheap. Similarly, in conditions of limited knowledge spillovers, capabilities to produce higher quality products correlate with the relative availability of specific factor supplies. Countries that are rich in those factors will thus have a cost and/or capability advantage in the production of quality.

The economic literature on vertical product differentiation has typically put much emphasis on the role of R&D. The main argument is that research, if successful, leads to the invention of new products or to major improvements in existing products (see, for instance, Stokey, 1988; Aghion and Howitt, 1994). Alternatively research may lead to the invention of new techniques that make it possible to produce higher quality goods. Two measures of R&D are available and have been used in empirical work: registered patents and R&D expenditure. The first can be seen as a measure of output of R&D and the second as a measure of input. Both measures have their drawbacks. R&D expenditure does not give any information about the success of the undertaken research. The problem with patents is that not every invention is patented and the propensity to patent is greater in some sectors than others, which would lead to a distorted picture. We will in our regressions alternatively use patents or R&D intensity as a measure for 'research and development' and it turns out that both measures lead to very similar results.

Early theoretical and empirical papers on vertical IIT have focused on the hypothesis that the production of high quality products is relatively capital intensive. High quality products will then be exported by countries that are relatively rich in physical capital. One problem with this hypothesis is that it would predict vertical IIT to be a phenomenon of trade between 'developed' and 'underdeveloped' countries, which are countries that differ a lot in their capital endowment. It has been pointed out before, however, that vertical IIT takes place among industrialized countries above all others (see the recent study by Fontagné and Freudenberg, 1997). These are countries that are relatively similar in their capital endowments. Indeed, first empirical tests on the role of physical capital (Hansson, 1994; Torstensson, 1996) find only weak support for the positive impact of this factor on product quality. Torstensson's analysis even seems to indicate that physical capital affects quality negatively, a result that we also find in our regressions.

Economists became particularly interested in the impact of workers' skills on product quality when they compared the trade performance of countries like the USA on the one hand and Japan and Germany on the other. The typical result of case-studies on the subject was that it is the high qualification of Japanese/German employees on the shop floor that allows firms in these countries to achieve a high output per worker and/or to produce a high quality good. In general, industrialized countries have very different education systems, which have an impact on the skill composition of their labour force. Also the willingness of firms to invest in training of their workers seems to vary substantially across countries. In our analysis we will try to see whether these differences have an impact on the quality of exports. We therefore included data on achieved education in some of our regressions. Comparable education data are, however, not available for many countries. This is one of the reasons why we used compensation rates and labour productivity as proxies for the skill composition of labour in other regressions.

So far we have only talked about factors that affect the cost structure or the capability of producing different quality levels. However, the characteristics of the markets to which producers supply their products also affect the rates of return obtained from producing and inventing high quality products. In this context the size, wealth and wealth distribution of/in the market they want to sell to will affect their decision. As we shall analyse the quality of imports to the European Union, the relevant market is that of the European Union. However, if access to this market is differentiated by producer (transport costs, entry barriers), then the size of the home market will also affect the decision of exporting firms as this is the first market they sell their products to. The size of the home market will therefore be included in some of the regressions as it may be expected that the very substantial investment outlays can be more easily recovered from a higher volume of sales.

In the following we will describe the results of several regressions performed with the three quality indicators presented in section 1. For each regression we will also give a more detailed description of the measures that have been used for the independent variables.

4.1 Quality Ranks

Table 3.7 shows the results of three regressions performed with our *quality rank* indicator. In all these regressions we used registered patents[11] as the measure for R&D. In the table this variable is referred to as *PAT*. The education data used are taken from OECD (1993) statistics and are classified according to the International Standard Classification of Education (ISCED). The variable $EDUC_{UPPERSEC}$ reflects the percentage of the workforce[12] that finished, at most, an upper secondary education, classified as ISCED 3. Upper secondary education covers in most countries the education between the age of 15 and 18/19. It includes among other things the vocational education in countries that have a so-called 'dual system'. Interestingly, the twenty OECD countries[13] included in the first regression differ much more in the percentage of population that finishes only secondary schooling than in the percentage of the population that continues studying and accomplishes a tertiary education. This last group is included in the regressions as the variable $EDUC_{TERTIARY}$. All the education data used in the regression refer to the year 1991. The variable *PHYSCAP* represents physical capital (per unit of output). We wanted this variable to enter as a capital stock and therefore calculated it as the accumulation of gross investment flows. The data for this variable have been taken from the OECD STAN statistics. The size of the exporting country has been measured by the national GDP and enters the regressions as the variable *GDP*. Finally a dummy $DUMMY_{EU}$ has been used in the regressions that include non-EU countries. With the aid of this dummy we want to control for the effect that EU membership may have on the quality rank variable.

Table 3.7 Quality ranks – regression results

	QR1	QR2	QR3
Constant (β_0)	–0.65**	–2.186**	0.81 **
	(–4.02**)	(–16.529)	(3.00)
PAT	0.029*	0.0368**	0.049**
	(1.93)	(12.79)	(3.80)
$EDUC_{UPPERSEC}$	0.030**	0.032**	1.736**
	(2.9)	(9.79)	(3.51)
$EDUC_{TERTIARY}$	0.020	0.010*	
	(1.109)	(1.959)	
PHYSCAP			0.000035 **
			(–10.77)
GDP	–0.0001	–0.000073 *	–0.00061 **
	(–0.69)	(–1.949)	(–2.92)
$DUMMY_{EU}$		–0.39**	
		(–5.189)	
R^2	0.683	0.079	0.126
Adjusted R^2	0.598	0.079	0.123
N	20	20441	973

Notes:
The values in brackets give White's heteroscedasticity consistent *t*-statistics.
* indicates a significance at the 10% level; ** a significance of 5%.
N gives the number of usable observations.

In regression QR1 we regressed national average ranks (QR^c, computed as explained in the first section) for the year 1992 on national values of the variables in consideration. We tested for the following functional specification, where the expected signs of the variables follow from the discussion in the above paragraph:

$$\text{(QR1)}\ QR^c = \beta_0 + \underset{(+)}{\beta_1 PAT^c} + \underset{(+)}{\beta_2\, EDUC^c{}_{UPPERSEC}} + \underset{(+)}{\beta_3\, EDUC^c{}_{TERTIARY}} + \underset{(+)}{\beta_4 GDP^c}$$

This regression includes the 20 OECD countries listed in note 13. The overall fit of the regression is quite high for a cross-country analysis with an R^2 of 0.683 and an adjusted R^2 of 0.598. Patents and upper secondary education turn out to be the only significant variables in the regression, with secondary education being significant at the 10 per cent level and patents at the 5 per cent level. Both variables also have the expected positive sign. Interestingly, tertiary education is insignificant though it has the expected positive sign. Country size instead has a negative sign, which goes against our expectations. But the variable is also insignificant. The results of regression QR1 thus give us reason to believe that R&D and upper secondary education are crucial factors in the determination of export quality. The weak point of this regression is, however, that it uses relatively few data points, namely 20. Unfortunately the availability of education data restricts severely the number of countries we can include in the analysis.[14] Nonetheless, to test the robustness of the obtained results, we performed regressions QR2 and QR3. In these regressions we refrained from computing average ranks and used instead individual product ranks $QR^{i,c}$ as the dependent variable. Regression QR2 does this for 1350 products and the 20 countries already included in regression QR1. This would give us a vector of 27 000 items if it were not for the missing values in a certain number of countries/products. The specification of the performed regression is the following:

$$\text{(QR2)}\ QR^{i,c} = \beta_0 + \underset{(+)}{\beta_1 PAT^c} + \underset{(+)}{\beta_2\, EDUC^c{}_{UPPERSEC}} + \underset{(+)}{\beta_3\, EDUC^c TERTIARY}$$

$$+ \underset{(+)}{\beta_4 GDP^c} + \underset{(-)}{\beta_5 DUMMY_{EU}}$$

Notice that the independent variables only vary per country, not per product, so that the level of variation of the dependent variable is considerably higher than in the independent variables. This probably explains for a great part the low R^2 found in regression QR2.[15] It is, however, encouraging to see that the results on the importance of the variables *PAT* and $EDUC_{UPPERSEC}$ do not change. R&D is only more significant in this regression. Tertiary education is now also significant, though only at the 10 per cent confidence level. Its coefficient is also smaller than the one for upper secondary education (0.01

compared to 0.03). GDP enters again with a negative sign and is significant this time at the 10 per cent level. The dummy has the expected negative sign and is significant. This probably reflects the effect of EU integration on product prices.

It has been mentioned before that the importance of the independent variable may vary depending on the industry we are looking at. It would therefore be interesting to perform the above regression at the industry level, to compare the results with the ones obtained using national data. This is done in regression QR3 that looks at the sector of electrical engineering. In this regression $QR^{i,c}$ refers to individual product ranks of 167 items belonging to the 2-digit industries SITC 76 and 77. We have education data on an industry level at our disposal only for a very restricted number of countries. Regression QR3 therefore only includes the countries France, the Netherlands, Germany, Italy, the UK and Denmark. We also could not include tertiary education in the regression, whereas data on gross investment were available for this industry and these countries. The functional specification tested thus amounts to the following equation, where the independent variables refer to data at the industrial level (apart from the values for national GDP, of course):

$$\text{(QR3)}\ QR^{i,c} = \beta_0 + \underset{(+)}{\beta_1 PAT^c} + \underset{(+)}{\beta_2 EDUC^c{}_{UPPERSEC}} + \underset{(+)}{\beta_3 PHYSCAP^c} + \underset{(+)}{\beta_4 GDP^c}$$

Because of the differences in the functional specification and in the countries included in the analysis, the results of QR3 and QR2 cannot be directly compared. Yet the results of QR3 give us useful indications concerning the robustness of the results obtained so far. Though still small the explanatory power is significantly higher in the regression at the industry level than in the one using national data. R&D and secondary education are again significant at the 5 per cent level, but this time this is also the case for GDP (country size) and physical capital. Yet these last two variables do not have the expected positive sign, but are negative instead!

Overall the three regressions performed with the quality rank variable allow for the following conclusions. R&D and upper secondary education seem to be important factors for attaining higher positions in the quality ranking. These two variables have a positive and significant impact in all the performed regressions. Also tertiary education seems to affect product quality positively, although its impact is not very significant in all the regressions. GDP and physical capital turn out to have an unexpected negative sign in some of our regressions. Our interpretation of this result is that it may reflect the presence of increasing returns to scale in the production of lower quality products. It is not unusual to observe that high quality products tend to be produced on a small scale, whereas lower quality products can be produced in bigger quantities. If the production of high quantities leads to lower average costs, this could explain the negative effect of the size of the home market on product quality. In the presence of economies of scale for lower quality products (and in the presence of transport costs or other barriers to trade), production will be located in countries with a large home market. If the production on a larger scale is also more capital intensive, this would explain the negative sign of our variable measuring physical capital. This, however, is only a first attempt to explain the obtained results and additional analyses would be needed to give a more robust explanation for the role of home country size and physical capital in the production of 'quality'.

4.2 Quality Segment and Price Quality Gap Indicators

In the regressions performed with the *quality segment* (*QS*) and the *price/quality gap* (*PG*) indicators[16] we used different measures for the factors that are expected to determine product quality than in the previous regressions:

- *Relative unit labour costs* is a cost variable but it also relates to the skill composition of the labour force and the amount of complementary inputs used in production (for instance, physical capital). To make this more explicit, this variable has been decomposed into *relative compensation rates per worker* (referred to as '*wages*' in short) and *(labour) productivity levels* (referred to as '*productivity*' and measured as output per employee). We expect that both higher compensation rates and higher labour productivity levels reflect (in longer-term equilibrium) a higher share of skilled workers in the labour force and/or a larger amount of complementary inputs used in production.
- R&D has this time been measured by *R&D intensities*, measured as a stock of R&D[17] per unit of constant price value added, and is expected to indicate differences in the degree to which different firms are capable to invent and/or produce higher quality products.

The above variables have always been constructed as 'relative' variables, that is, a country's labour compensation rates, productivity levels and R&D intensities relative to the group of competitors. In this context the competitors' values were calculated as weighted averages, with the competitors' weights being their export shares in the particular industry. To derive relative variables they had to be compared in a common currency. For relative wage rates (WR), current exchange rates were used while relative productivity levels (PR) were compared in purchasing power parity (PPP) rates. Relative labour unit costs were then calculated as the ratio WR/PR thus defined.

Information for these explanatory variables was extracted from the OECD's STAN database. This database reports information only at a much higher level of aggregation than the NACE 3-digit level on the basis of which the *QS* and *PG* indicators were calculated. Hence these indicators were aggregated to a 2-digit NACE level which coincided with the STAN type of classification. Our panel therefore includes data for three 2-digit engineering branches (NACE 32, 34, 37) over the period 1988–94. The countries included in the analysis differ according to the regressions, depending on whether the R&D intensity variable was included or not (this variable could only be constructed for a subset of OECD countries).[18]

In the following we introduce the functional form of the performed regressions and discuss shortly our prior expectations regarding the way our regressors would affect the two different dependent variables (*PG* and *QS*). We should emphasize some of the differences between the two quality indicators in this regard. The *QS* indicator is based upon a comparison of the commodity composition of countries' exports, while the *PG* indicator is directly based on prices. In the first place we expect *PG* to be more sensitive to short-run cost pressures such as exchange rate fluctuations or wage pressures. In the second place the role of R&D and productivity may change according to the dependent variable. Whereas we have so far emphasized the importance of R&D for product innovation (affecting a producer's product composition and hence QS^I directly), it is also directed towards process innovation. Whereas improvements in the process technology should therefore (in theory) not affect *QS* they will lead to a reduction in unit costs and hence a fall in the relative price. Hence, while product innovation affects *PG* positively, we expect process innovation to have a negative impact on it (through cost reductions), so that the sign for the R&D variable in the price gap regressions might take a positive or negative value and/or the coefficient might turn out to be insignificant. A similar indeterminacy exists with the productivity variable: it is perfectly possible for two producers to have different levels of (labour) productivity but the same levels of unit costs reflecting different positions on the same iso-cost curve (that is, different amounts of complementary inputs combined with a higher or lower level of labour inputs). The sign for the productivity variable in the *PG* regressions is thus also ambiguous a priori.

The above considerations lead us to use the following specification[19] for the regression explaining a country's positioning in the high quality segment:

$$QS^I = \beta_0 + \underset{(+)}{\beta_1\, wages_{ex}} + \underset{(+)}{\beta 2\, prod_{pp}} + \underset{(+)}{\beta_3\, R\&D_{pp}} + D_c + D_j$$

As the dependent variable in this regression we used a country's relative representation in the highest quality segment (QS^I). The subscripts *ex* and *pp* indicate whether the variables are expressed in current exchange rates or in purchasing power parities. The D_c and D_j denote country or country group dummies and industry dummies respectively. The fact that the compensation rate variable has been measured in current exchange rates implies that it combines longer-run determinants of compensation rates (such as skill composition) with short-run cost pressures which could result from a temporary overvaluation of the exchange rate or from other short-term pressures related to wage rates. However, both of these factors, improvements in the skill composition of the labour force, and pressures of higher wage rates, should affect the dependent variable in the same direction: one acts as a facilitator and the other as a cost pressure to improve relative product quality.

Between the first two variables, wage rates and productivity levels, one can expect a close relationship and we have used an instrumental variables[20] approach to deal with the potential interrelationship between them. As this approach, however, did not produce satisfactory results, we adopted an alternative approach and omitted the less significant[21] variable, which turned out to be '*wages*' in this case. The results of this latter regression are the ones reported in Table 3.8.

Table 3.8 Quality segments and price gaps – regression results

Dependent variable	QS1[1]	QS1[1]	PG[2]	PG[2]
Constant	–1.761**	–1.886**	0.096**	0.147**
	(11.481 **)	(13.72)	(9.41)	(11.19)
Wage rates	–0.369*			
	(1.81)			
Productivity levels	0.548**	0.375 **		
	(3.60)	(3.14)		
Labour unit costs			0.330**	
			(7.29)	
R&D intensities	0.324**	0.263 **		0.214**
	(5.06)	(4.82)		(11.19)
Industry dummy 1	0.184**	0.189**		
	(2.64)	(2.71)		
Industry dummy 2	–0.123 *	–0.113 *		
	(1.77)	(1.61)		
R^2	0.194	0.183	0.159	0.435
Adjusted R^2	0.177	0.170	0.156	0.430
N	243	243	282	243

Notes:
The values in brackets give White's heteroscedasticity consistent *t*-statistics.
* indicates a significance at the 10% level; ** a significance of 5%.
N gives the number of usable observations.
[1] Dependent variable in logs.
[2] All variables except dummies in logs.

The specification of the price gap regression was the following.[22]

$$PG = \beta_0 + \underset{(+)}{\beta_1\, wages_{pp}} + \underset{(?)}{\beta_2\, prod_{pp}} + \underset{(?)}{\beta_3\, R\&D_{pp}} + D_c + D_j$$

The question marks below the regressors *productivity* and *R&D* reflect the considerations made before that we cannot predict in which direction these two variables will affect relative price gaps.

To deal with the possible endogeneity of wages and productivity, we report here the regressions with the combined labour unit cost variable. When the two variables were both included the wage variable was the more significant of the two, confirming our prior expectation that current exchange rate and wage pressures affect the price variable more than the *QS* variable. Omitting Austria, Belgium, Portugal and Greece from the country sample allowed us to include the *R&D* variable which turned out to be highly significant.

Overall, our results show that R&D intensities and relative productivity levels turned out to affect QS^I positively and significantly. Relative R&D intensities were also very effective in explaining vertical price differentiation across suppliers.

6 CONCLUSIONS

This chapter has examined the evidence for vertical product differentiation in EU markets. More specifically, it has looked at trade (including intra-EU) in EU markets. The focus on trade allowed us to use extremely disaggregated product (volume and price) statistics and to construct a number of measures of 'vertical' or 'quality' differentiation in EU markets. We were particularly interested in analysing the determinants of the quality positioning of the different national producers (EU and non-EU) in the product markets of the EU. The following are the principal results from our analysis:

- There is ample evidence of very marked vertical product differentiation in EU markets.
- The 'quality' positioning of different national producers is, at first approximation, linked to an economy's developmental level. Hence more advanced economies (in terms of GDP per capita) also rank in higher quality positions. The analysis showed evidence of a strong hierarchy of the (average) national suppliers to EU markets and this hierarchy is also evident across the spectrum of more advanced ('northern') EU economies and the less advanced ('southern') economies.

- On closer inspection, the econometric analysis found consistent evidence for skill and R&D variables to determine a country's quality positioning. There is also evidence of a link between high wage rates, high productivity and high quality positioning, although the directions of causation have not been determined in the analysis.
- For an assessment of the relative competitive positions of EU vs non-EU advanced economies the analysis is not conclusive, as EU economies have the benefit of easier access to other EU markets as compared to non-EU exporters, and hence the EU trade figures are likely to represent a wider range of producers (including a greater proportion of less efficient and lower quality producers) than do the trade flows involving non-EU trading partners. This conjecture has been borne out by a separate analysis of bilateral trade flows between EU economies and the USA in which the price/quality gaps between (the average) EU supplier and the US supplier (which had originally been found in EU trade) disappeared.
- One of the most interesting observations regarding the European economy, is the evidence concerning the range of high to low quality producers found in Europe and that this range has been substantially extended through the entry of East European producers. These entered, after trade liberalization, at the lowest end of the quality spectrum (in the industries we examined), at levels similar to those of the lowest quality producers worldwide. However, there is evidence of strong upward movements of the 'western CEECs' (particularly Hungary, the Czech Republic and Slovenia) while there is evidence for stagnation or falling behind for the 'eastern CEECs'.
- From the policy point of view, the econometric results concerning the effectiveness of skill and R&D variables affecting the quality positioning of producers and the further links to wage and productivity levels are in line with other research results. These emphasize that high levels of income stem from successful efforts made in the areas of skill upgrading and technology development, leading to high rates of product innovation and the positioning of producers in the high quality segments of vertically differentiated product markets.

NOTES

1. Throughout the chapter the data used are extracted from EUROSTAT trade statistics. Data on prices of traded goods are not directly available, but data on values of imports are and so are data on quantities and/or weights of the imports. It is therefore possible to compute 'values per item' and 'values per ton'. The latter has been used throughout the analysis, again for reasons of data availability. See Jansen (1998) and Aiginger (1996) for a discussion of the advantages and disadvantages of the use of this measure.

2. Hufbauer (1970) uses the coefficient of variation of product prices as a measure of the degree of vertical product differentiation. Hansson (1994) and Torstensson (1996) use product prices in an exercise related to ours.
3. Depending on the case we used data at the 8-digit level of the Combined Nomenclatura (CN) or 5-digit SITC.
4. See Fersterer and Landesmann (1998) for a more detailed description of the *PG* and the *QS* indicators; the results presented in this paper on these two types of indicators come from joint work with Josef Fersterer and Johann Burgstaller, both at the Johannes Kepler University in Linz.
5. EU here refers to the EU 12 market.
6. The share of each quality segment amounts – by construction – exactly to one third of total EU imports.
7. *J* is the total set of industries included in the analysis. In this case these are the following 18 SITC groups: 11, 12, 51, 52, 54, 61, 64, 65, 71–78, 84, 85.
8. Another reason for using the QR indicator is that data at a higher level of aggregation were used for its construction (5-digit as compared to 8-digit in the case of the *PG* and *QS* indicators); consequently price comparisons were made on less narrowly defined product categories which would make the inference from cardinal price to quality ranking less reliable.
9. These include NACE 321–328 (mechanical engineering), NACE 330 (office machinery), NACE 341–347 (electrical goods) and NACE 371–374 (instruments). The transport equipment industries were excluded from this sample because of the dominance of single product items in the trade structures of some of the branches.
10. The specifications of these regressions amount to $LPG^c_{jt} = \alpha^c_t \cdot dummy^c + \varepsilon^c_j$ which were estimated over countries (or country groups) *c* and across industries *j* belonging to a particular industry group (such as the engineering industries) and for time periods t = 1988–90 or 1992–94 or 1995–96 (that is, three- or two-year averages). *PG* refers to the dependent variable, the price/quality gap variable; ε^c_j refers to the usual randomly distributed stochastic term; see Tables 3.2 and 3.3 for the results for two groups of industries, the engineering industries and textiles, clothing and leather products. The same specification was used further on for the QS^I quality segmentation variables. At the bottom of the two tables are estimates for country groups, which comprised: EUN (northern countries of the European Union except Ireland but including Italy), EUS (southern EU countries – Spain, Portugal and Greece), EFTA (with Switzerland, Austria, Sweden and Finland), CEEC1 (a group of 'western' CEECs – CSFR/the Czech Republic, Hungary, Poland, Yugoslavia/Slovenia), CEEC2 (a group of 'eastern' CEECs – Bulgaria, Romania, Slovakia for the later two periods and the Soviet Union/Russia) and the NICs (NICs 1 comprising Hong Kong, Singapore, South Korea and Taiwan; NICs 2 comprising Indonesia, Malaysia, the Philippines and Thailand).
11. The data used have been taken from statistics of the US Patent Office. To correct for country size the number of patents has been divided by GDP. Besides this the values used are the patents obtained in the years 1986–92, to avoid distortions due to exceptionally successful years.
12. To be more precise, the variable refers to the percentage of the population 25 to 64 years of age, whose highest completed level of education is an upper secondary education.
13. These 20 countries are France, Belgium, Netherlands, Germany, Italy, the United Kingdom, Ireland, Denmark, Portugal, Spain, Norway, Sweden, Finland, Switzerland, Austria, Turkey, the United States, Canada, Australia, New Zealand.
14. At the same time, however, it is useful to include only OECD countries in the analysis, as this allows us to see whether our variables can explain different export behaviour of relatively similar countries.
15. As the dependent variable consists of discrete and ranked numbers it would be more correct to use an ordered probit regression instead of the least squares method used in QR2. The results on significance and sign of the different variables should, however, not change significantly using the alternative methodology.
16. The regressions reported in this section rely on joint work with Josef Fersterer of the Department of Economics at the Johannes Kepler University, Linz.
17. The R&D ('knowledge') stock has been constructed from R&D expenditure flow data by perpetual inventory method with an assumed linear yearly depreciation rate of 0.2 (see also Coe and Helpman, 1995; Landesmann and Pfaffermayr, 1997).

18. Sample 1 includes Germany, France, Belgium, the Netherlands, Italy, the United Kingdom, Austria, Finland, Sweden, Denmark, Spain, Portugal, Greece, the United States, Japan and Canada. Sample 2 includes the same countries apart from Belgium, Austria, Portugal and Greece.
19. Note that Table 3.8 reports the result for this specification with the dependent variable in logs. We also experimented with a version with all variables except dummies in logs, and the results were very similar. For details see Fersterer and Landesmann (1998).
20. The additional instruments used were hours of work lost through strike activity and price–cost margins.
21. In the regression using both variables, wages and productivity, wages turned out to have an unexpected negative sign, but was not significant, whereas productivity was significant and positive. See Fersterer and Landesmann (1998) for more details.
22. Note that Table 3.8 reports the result for this specification with all variables (except dummies) in logs.

REFERENCES

Aghion, P. and P. Howitt (1994), 'Growth and Unemployment', *Review of Economic Studies*, pp. 477–94.

Aiginger, K. (1996), 'Unit Values, Quality Ladders and Non Price Competition', paper prepared for the 'Conference on New Indicators for the Knowledge-Based Economy', OECD, June.

Cabrales, A. and M. Motta (1996), 'Country Asymmetries, Endogenous Product Choice and the Speed of Trade Liberalization', *CEPR Discussion Paper* No. 1326.

Coe, D.T. and E. Helpman (1995), 'International R&D Spillovers', *European Economic Review*, **39** (5), pp. 859–87.

Fersterer, J. and M. Landesmann (1998), 'Vertical Product Differentiation in International Trade: An Econometric Investigation', *Working Paper* No. 8, The Vienna Institute for International Economic Studies, WIIW, Vienna.

Fontagné, L. and M. Freudenberg (1997), 'Intra-Industry Trade: Methodological Issues Reconsidered', CEPII, Document de travail 97-1, Paris.

Greenaway, D., R. Hine and C. Milner (1994), 'Country-Specific Factors and the Pattern of Horizontal and Vertical Intra-Industry Trade in the UK', *Weltwirtschaftliches Archiv*, Band 130, 77–100.

Hansson, P. (1994), 'Product Quality and Vertical Product Differentiation as Determinants of Intra-Industry Trade in Swedish Manufacturing', *Working Paper* No. 122, Trade Union Institute for Economic Research (FIEF).

Hufbauer, G.C. (1970), 'The Impact of National Characteristics and Technology on the Commodity Composition of Trade in Manufactured Goods', in R. Vernon (ed.), *The Technology Factor in International Trade*, New York: Colombia University Press.

Jansen, M. (1998), 'How to Compete in Quality: Factors Explaining Vertical Intra-Industry Trade', mimeo, Barcelona: University Pompeu Fabra.

Landesmann, M. and J. Burgstaller (1998), 'Trade Performance of East European Producers on the EU Market', *Research Report* 255, Vienna: The Vienna Institute for International Economic Studies (WIIW).

Landesmann, M. and M. Pfaffermayr (1997), 'Technological Competition and International Trade', *Applied Economics*, **29**, 179–96.

Stokey, N. (1988), 'Learning by Doing and the Introduction of New Goods', *Journal of Political Economy*, **96**, 701–17.

Torstensson, J. (1996), 'Can Factor Proportions explain Vertical Intra-Industry Trade?', *Applied Economic Letters*, forthcoming.

4. Employment dynamics and structural change in Europe

Mario Pianta and Marco Vivarelli

1 INTRODUCTION

In Europe, since the recession of the early 1990s, a combination of macroeconomic and structural factors has led to an unprecedented 'divorce' between the growth of production and the stagnation (if not a continuing fall) in the number of jobs. There are signs that the present forms of structural and technological change and the pressure imposed by growing international competition have weakened – or even eliminated – the association between growth and employment which has been one of the main features of the model of growth of postwar decades. There is now a growing concern about 'jobless growth' as a source of social exclusion.

In this chapter, comparing European economies with the United States and Japan, we investigate how in an era of economic integration, European competitiveness and specialization patterns have led to a growth regime that has been particularly 'unfriendly' to employment. The poor European performance in terms of employment is then related to structural and technological factors, pointing out the importance of the sectoral composition of industry and the complex role played by technological change.

This analysis of the employment effects of structural change and integration in Europe is clearly in line with the overall approach adopted in this volume, but it is also at odds with the usual approach to the issue of unemployment found in the literature. More specifically, the focus on the importance of structural and technological change in the employment outcomes challenges the conventional wisdom that confines the explanations of European unemployment to labour market factors, such as labour costs (higher minimum wages and upward wage dynamics) or rigidities in terms of hiring/firing rules and the welfare and benefit systems.

In order to introduce the main arguments on the relation between technological and structural factors and employment, the next section will present a review of the theories on technological change and employment. In section 3, the long-term evolution of employment in Europe, Japan and the USA will be compared. Section 4 will examine the role of the sectoral structure of industry and of technology in affecting employment. Finally, section 5 will sum up the main findings of this study.

2 TECHNOLOGICAL CHANGE AND EMPLOYMENT

Is there a relationship between the sectoral structure of a given economy, technological change and the demand for labour (Ricardian unemployment) or is unemployment entirely determined by aggregate demand (Keynesian unemployment) and wage dynamics (classical unemployment)? Mainstream economics tends to neglect the possibility of structural unemployment, with technological change generally taken as exogenous and unemployment generally studied as a short-term phenomenon rather than a structural problem. In this section this conventional wisdom will be challenged and the theory about the link between technological change and employment will be briefly discussed (for broader discussions of this complex topic, see Petit, 1995 and Vivarelli, 1995).

From the very beginning of economics as a discipline, it has tried to dispel all concerns about the harmful effects of technological 'progress'. In particular, economists put forward a theory that Marx later called the 'compensation theory' (see Marx, 1961, Vol. 1, Chapter 13 and 1969, Chapter 18). This theory argues that technological change, besides saving labour, triggers different market compensation mechanisms which can counterbalance the initial labour-saving impact of process innovation (for an extensive analysis, see also Vivarelli, 1995). These compensation mechanisms run as follows.

First, the same process innovations that displace workers in the user industries create new jobs in the capital sectors where the new machines are produced. Second, while process innovations displace workers, these innovations themselves lead to a decrease in the unit costs of production and (in a competitive market) this effect is translated into decreasing prices. These, in turn, stimulate demand, production and employment (see Stoneman, 1983; Vivarelli, 1995). Third, in a world where competitive convergence is not instantaneous, innovative entrepreneurs may accumulate extra profits. These profits are invested and new jobs are created. Fourth, as with other forms of unemployment, technological unemployment should lead to a fall in wages and this should lead to an increase in the demand for labour.

In addition to these four compensation mechanisms, many economists have underlined that besides process innovation, technological change may also take the form of product innovation (introduction of new products, or improvement of existing products). In this case, the labour-saving effects of technology are not obvious. In fact, product innovations may have a positive impact on employment since they may have a positive impact on demand (see Freeman, Clark and Soete, 1982; Freeman and Soete, 1987, 1994; Vivarelli, 1995).

However, some important critiques can be opposed to each of the four compensation mechanisms. The compensation mechanism 'via new machines' cannot fully counterbalance the initial job losses since machines are profitable only when they are built with less labour than they replace. Moreover, labour-saving technologies affect the capital goods sector as well. Also, the new machines can be implemented simply by scrapping. If this is the case, there is no compensation at all (see Freeman, Clark and Soete, 1982, p. 160). The mechanism via a decrease in prices relies on Say's law and does not take into account either that demand constraints might occur or that the demand elasticity may be lower than one. In addition, the effectiveness of this mechanism depends on the hypothesis of perfect competition (see Sylos Labini, 1969, p. 160). The compensation mechanism 'via new investments' also relies on Say's law: cumulated profits due to technological change must be entirely and immediately translated into additional investments. Moreover, the intrinsic nature of the new investments matters. If these are capital-intensive and labour-saving, compensation can only be partial. As far as the mechanism 'via decrease in wages' is concerned, it has to be underlined that, besides inducing firms to hire additional workers, this also leads to a decrease in aggregate demand. In addition, with cumulative and localized technological change, the hypothesis of perfect substitutability between capital and labour assumed by this mechanism appears to be quite arbitrary (see Rosenberg, 1976; Dosi, 1988; Freeman and Soete, 1987, p. 46).

Thus, on the whole, the different mechanisms of the compensation theory form a very complex picture where there are a lot of ways for counterbalancing the direct labour-saving effect, but also a lot of possible drawbacks and hindrances (see Pasinetti, 1981, pp. 90ff.). Thus, one is tempted to conclude that new products are still the more powerful way to counterbalance labour-saving process innovations. However, the employment creation potential of product innovations is by no means automatically realized. Countries may be characterized by different technological trajectories, different degrees of market competition, different investment functions, different labour market adjustments, or different specialization patterns. All these factors have a direct impact on the role of labour-saving technological change and on the effectiveness of different compensation mechanisms. The remainder of this chapter will try to quantify some of these factors for the European case.

3 LONG-RUN EMPIRICAL EVIDENCE

An overview of the dynamics of economic and employment growth in the long run is a necessary first step in order to set the context for the analysis of the technological and structural factors and their employment impact in the European integration process. The aggregate empirical analysis is based on data for the USA, Japan and the 15 EU countries over the period 1975–96. The link between growth in value added and employment will be investigated both in aggregate and separately for industry and services.

As can be seen from Figure 4.1, over the period 1975–95 EU employment has been stagnating (growth of about 6 per cent) while the USA shows an overall increase of about 45 per cent and Japan of about 23 per cent. In other words, over the long run, Europe has been characterized by almost jobless growth, while the USA and Japan have exhibited a clearly superior job-creation capacity. It is important to notice that these differences would probably turn out even larger if total working time evolution – and not the number of employees – was taken into account.[1]

These differences are even more marked if attention is turned towards the industrial sector alone. As it can be seen in Figure 4.2, while the USA and Japan are characterized by (moderately) positive trends (about +15 per cent), Europe exhibits a clear downward pattern (about –18 per cent).

Even in services, as shown in Figure 4.3, although all the three areas show steadily increasing trends, the USA is characterized by the best employment performance (about +63 per cent), followed by Japan (about +46 per cent) and the EU (about +42 per cent).

Thus the long-run capacity to couple economic growth with an increase in the use of labour seems to be confined to the extra-European countries. This overall effect is due to European jobless growth in the industrial sector and to slower job expansion in services. Of course, the weak employment intensity of European growth might be considered as the result of a stronger productivity performance, which might improve competitiveness and lead to faster growth. However, this does not seem to be the case in the EU, where GDP growth is indeed lower than in the USA and Japan. So, in terms of employment, job losses due to productivity gains are not compensated through a competitiveness effect.

This picture is confirmed by the data reported in Tables 4.1, 4.2 and 4.3. In these tables the overall variations in real value added and employment are reported for the economy as a whole, and separately for industry and services. In the last column the employment elasticity is reported (computed as the ratio of the two previous columns).

As far as the whole economy is concerned, the values in the third column of Table 4.1 clearly show the higher labour intensiveness of US economic growth in comparison with Europe. There is a clear ranking, with all the European countries showing an elasticity lower than that of the Japanese and much lower than the US (the exceptions being three small countries, Austria, Greece and the Netherlands; in the latter the expansion in the number of employees is mainly due to a dramatic increase of part-time jobs). Looking at the four largest

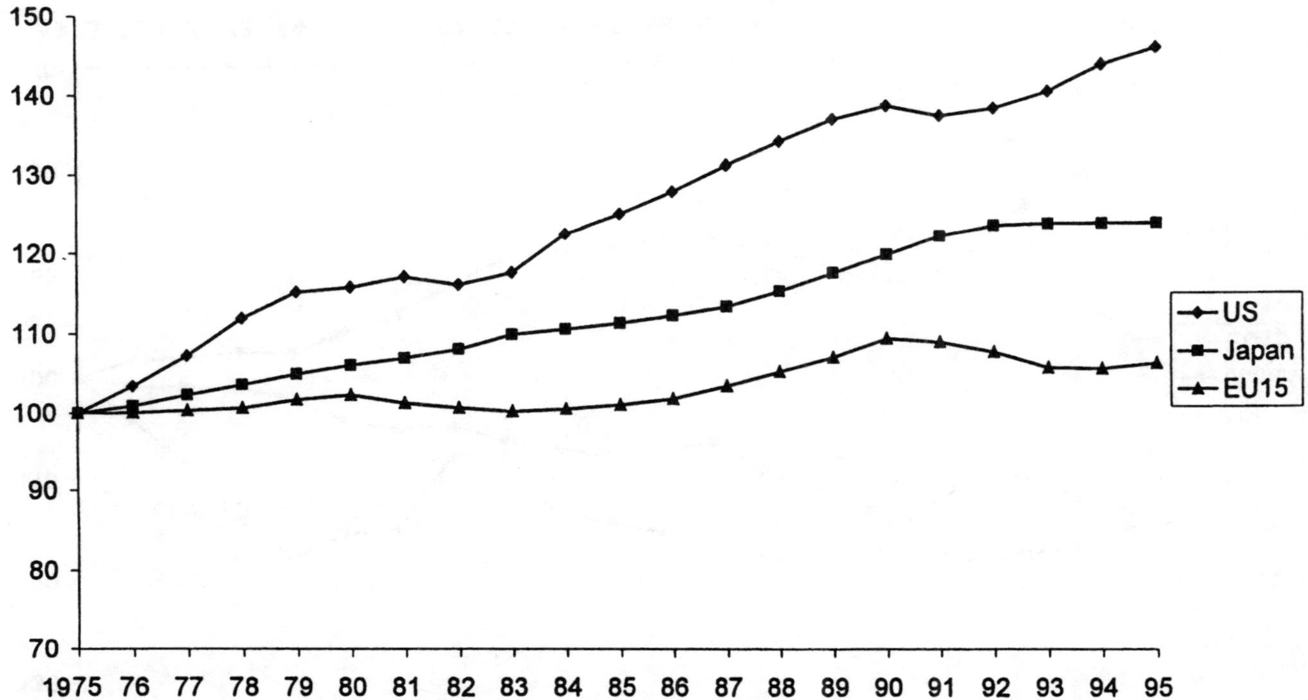

Figure 4.1 *Total employment*

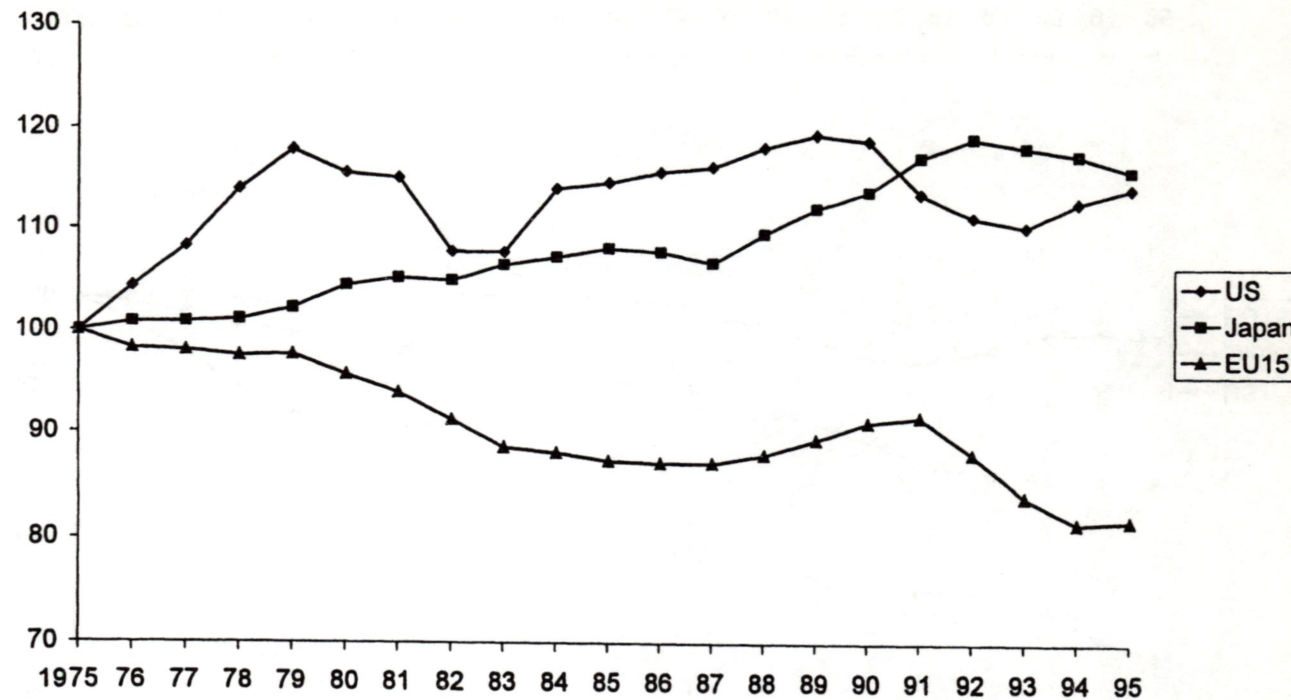

Figure 4.2 *Employment in industry*

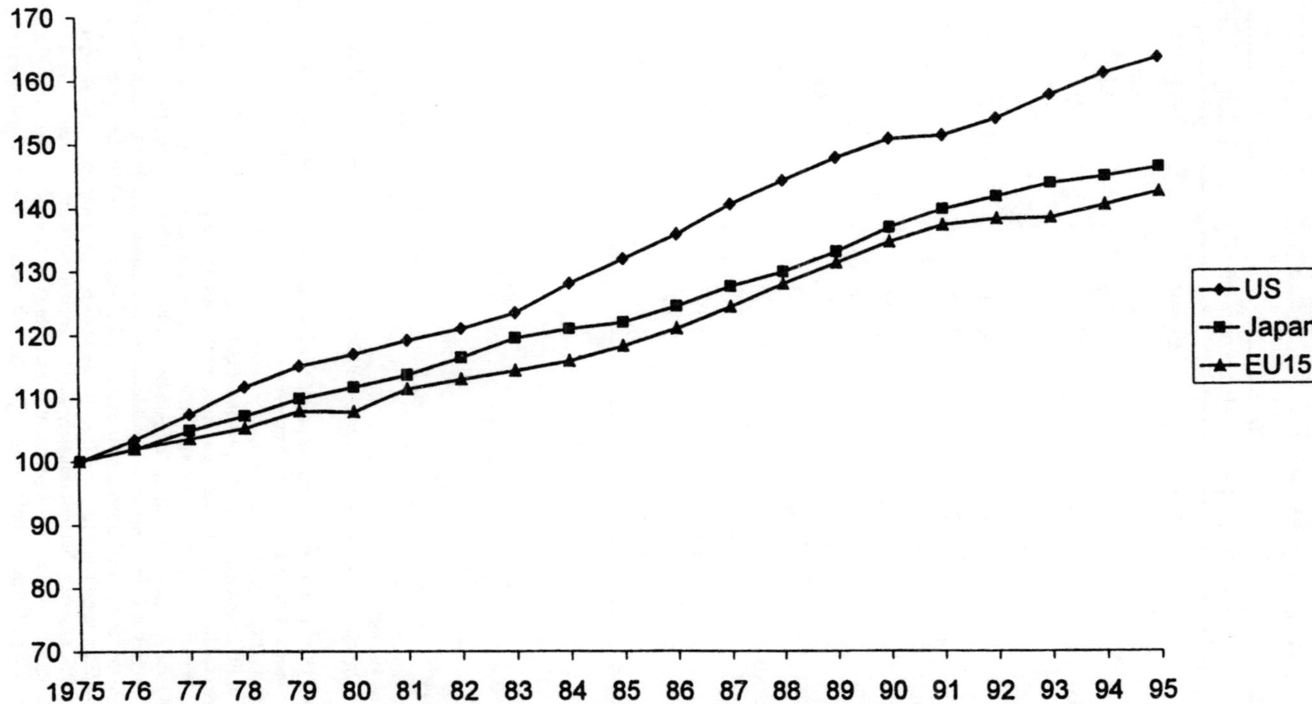

Figure 4.3 Employment in services

Table 4.1 Growth and employment in the whole economy. Long-run percentage changes, 1975–96

Countries	Real GDP	Employment (E)	Elasticity (E/GDP)
US	78.82[a]	47.60[a]	0.60[a]
Japan	85.92[a]	24.21[a]	0.28[a]
France	53.87[a]	5.21[a]	0.10[a]
Germany	46.65[b]	9.39[b]	0.20[b]
Italy	65.09[a]	2.80[a]	0.04[a]
UK	55.16[a]	5.54[a]	0.10[a]
EU4	58.64[a]	4.33[a]	0.07[a]
Austria	55.73[c]	27.02[c]	0.48[c]
Belgium	42.17[d]	0.79[d]	0.02[d]
Denmark	56.71[a]	11.19[a]	0.20[a]
Finland	57.00[a]	–5.60[a]	–0.10[a]
Greece	61.25[e]	19.84[e]	0.32[e]
Ireland	157.58[a]	23.19[a]	0.15[a]
Netherlands	61.98[a]	51.47[a]	0.83[a]
Portugal	88.86[a]	20.17[a]	0.23[a]
Spain	61.46[a]	–0.39[a]	–0.01[a]
Sweden	35.13[a]	–2.44[a]	–0.07[a]
EU15	56.24[e]	6.14[e]	0.10[e]

Notes:
Time interval:
a. 1975–96
b. 1975–90
c. 1975–94
d. 1976–93
e. 1975–95

Sources and methods:
Real GDP: *Data Stream* for the USA, Japan, the UK; *OECD National Accounts* for all the other countries.
Data for EU15 and EU4 are also based on *OECD National Accounts* and are expressed in US dollars at constant 1990 prices and exchange rates.
Employment: *OECD Quarterly Labour Force Statistics* for every country.
The series of EU15 and EU4 are also based on *OECD Quarterly Labour Force Statistics*; since this source presented lack of data for some countries for the most recent years, we had to consider other databases. Particularly, we derived the necessary information to complete the series from *EUROSTAT* for Austria (1995 only), Belgium (from 1993), and France (from 1990).
To obtain a series as homogeneous as possible, we had to subtract the East German employment figures from the total after German reunification. Therefore, we have calculated the employment rate of change for West Germany only from the *Data Stream* data set, and we have applied these rates to the *OECD series*, that unfortunately did not report the statistics relative to West Germany from 1990 on. This operation is equivalent to assuming that from the date of reunification the proportion of West German employment to East German employment remained constant.

European countries, West Germany shows the best employment trend (although relative to the pre-unification period and at a rate lower than the Japanese and US ones), while the other three countries show elasticities either equal to or lower than 0.10 (against 0.60 in the USA and 0.28 in Japan). The EU average is about 0.10, thus confirming a picture in which – with very few exceptions – European countries have enormous difficulties in translating economic growth into jobs.

Table 4.2 Growth and employment in industry. Long-run percentage changes, 1975–96

Countries	Real value added (RVA)	Employment (E)	Elasticity (E/RVA)
US	26.27[a]	15.20[g]	0.39[a]
Japan	81.05[a]	15.22[g]	0.22[a]
France	13.03[b]	–26.18[b]	–2.01[b]
Germany	14.63[d]	–4.08[c]	–0.13[c]
Italy	15.05[d]	–14.55[d]	–0.97[d]
UK	6.09[a]	–31.59[g]	–4.13[a]
EU4	11.97[a]	–19.17[b]	–1.37[a]
Austria	32.58[a]	4.24[a]	0.13[a]
Belgium	61.79[c]	–31.50[e]	–0.44[c]
Denmark	22.56[a]	–13.15[g]	–0.44[a]
Finland	5.45[a]	–25.10[g]	–5.75[a]
Greece	20.87[a]	0.78[a]	0.04[a]
Ireland	112.38[e]	–5.04[e]	–0.04[e]
Netherlands	9.62[f]	0.58[f]	0.06[f]
Portugal	75.61[e]	23.39[g]	0.21[e]
Spain	34.61[g]	–20.60[g]	–0.60[g]
Sweden	–7.74[a]	–27.54[g]	–4.13[a]
EU 15	13.22[a]	–18.28[b]	–1.22[a]

Notes:
Industry is defined as the Major Divisions 2, 3, 4 and 5 of the ISIC.
Time interval:
a. 1975–93
b. 1975–95
c. 1975–90
d. 1976–96
e. 1975–92
f. 1977–96
g. 1975–96

Sources and methods:
RVA: *OECD Quarterly National Accounts* for France, Germany, Italy, the Netherlands, Spain; UN *National Accounts Statistics: Analysis of the Main Aggregates* for the other countries.
Employment: see Table 4.1.

Table 4.3 Growth and employment in services. Long-run percentage changes, 1975–96

Countries	Real value added (RVA)	Employment (E)	Elasticity (E/RVA)
US	96.84[a]	65.79[g]	0.59[a]
Japan	111.23[a]	47.62[g]	0.39[a]
France	90.71[b]	40.98[b]	0.45[b]
Germany	100.74[d]	29.73[c]	0.43[c]
Italy	94.43[d]	37.80[d]	0.40[d]
UK	77.88[a]	31.91[g]	0.31[a]
EU4	86.68[a]	37.18[b]	0.39[a]
Austria	108.36[a]	51.35[a]	0.47[a]
Belgium	94.94[c]	25.40[e]	0.25[c]
Denmark	49.53[a]	30.51[g]	0.55[a]
Finland	63.25[a]	25.65[g]	0.33[a]
Greece	65.71[a]	82.94[b]	1.10[a]
Ireland	128.79[e]	67.49[g]	0.30[e]
Netherlands	74.56[f]	77.65[f]	1.04[f]
Portugal	192.23[e]	110.32[g]	0.54[e]
Spain	142.27[g]	55.11[g]	0.39[g]
Sweden	65.04[a]	21.29[g]	0.33[a]
EU15	90.47[a]	41.86[b]	0.42[a]

Notes:
Services are defined as the Major Divisions 6, 7, 8 and 9 of the ISIC.
Time interval:
a. 1975–93
b. 1975–95
c. 1975–90
d. 1976–96
e. 1975–92
f 1977–96
g. 1975–96

Sources and methods: see Table 4.2.

In the industrial sector, employment stagnation in Europe turns into pure jobless growth. Apart from Portugal, Austria and the Netherlands, all European countries exhibit negative trends that are particularly severe in the UK (–32 per cent), Belgium (–31 per cent up to 1992), Sweden (–28 per cent) and France (–26 per cent up to 1995). This leads to an EU average equal to –18 per cent (to 1995), which contrasts with a positive pattern (+15 per cent) both in Japan and the USA. Thus, while the USA and Japan keep on showing positive employment elasticities even in the industrial sector, European countries are characterized by a clear tendency towards deindustrialization (in terms of employment).

Turning the attention to services, all countries show positive variations and elasticities, but most European countries still lag behind the USA and Japan in terms of absolute variations. In terms of elasticities, the USA is still leading, while Japan drops down to the European average.

In summary, the USA and, to a lesser extent, Japan show overall increases in the use of labour which are based on positive job creation in industry and by higher employment elasticities in services. European countries are characterized by an overall employment stagnation that is mainly due to dramatic labour saving in the industrial sectors. In other words, in Europe the service sector hardly compensates the job losses of the industrial sector, while in the USA and Japan the positive employment performances are determined by both service expansion and positive long-run employment elasticities in industry.[2] Thus the European labour-saving bias in the industrial sectors plays an important role in determining the overall poor European employment performance. The discussion in the next section will be devoted to the composition of European industry in order to understand the structural reasons underlying its revealed long-term tendency in favour of labour-saving restructuring.

4 THE ROLE OF THE INDUSTRIAL STRUCTURE AND TECHNOLOGICAL CHANGE

This section addresses two issues. First, the importance of the *composition effect*, showing how the slow growth or decline of jobs in Europe is rooted in an industrial structure where sectors with the worst dynamics are over-represented. Second, *the direct impact of technological change* at the sectoral level will be examined, summarizing various research results on the job-creating or job-replacing patterns in these countries. The data used in the analysis of the first issue come from the OECD STAN database (1996) for 31 manufacturing sectors (ISIC classes). The USA, Japan and all countries of the European Union (with the exception of Ireland and Luxembourg and with the inclusion of Norway) are considered. The variables looked at include value added (in constant prices) and employment (number of persons engaged), measured as average annual rates of change from 1975 to 1994.

Figure 4.4 shows for the OECD aggregate the rates of change of value added and employment of the 31 manufacturing sectors, illustrating the overall pattern of structural change which has taken place over the two decades investigated

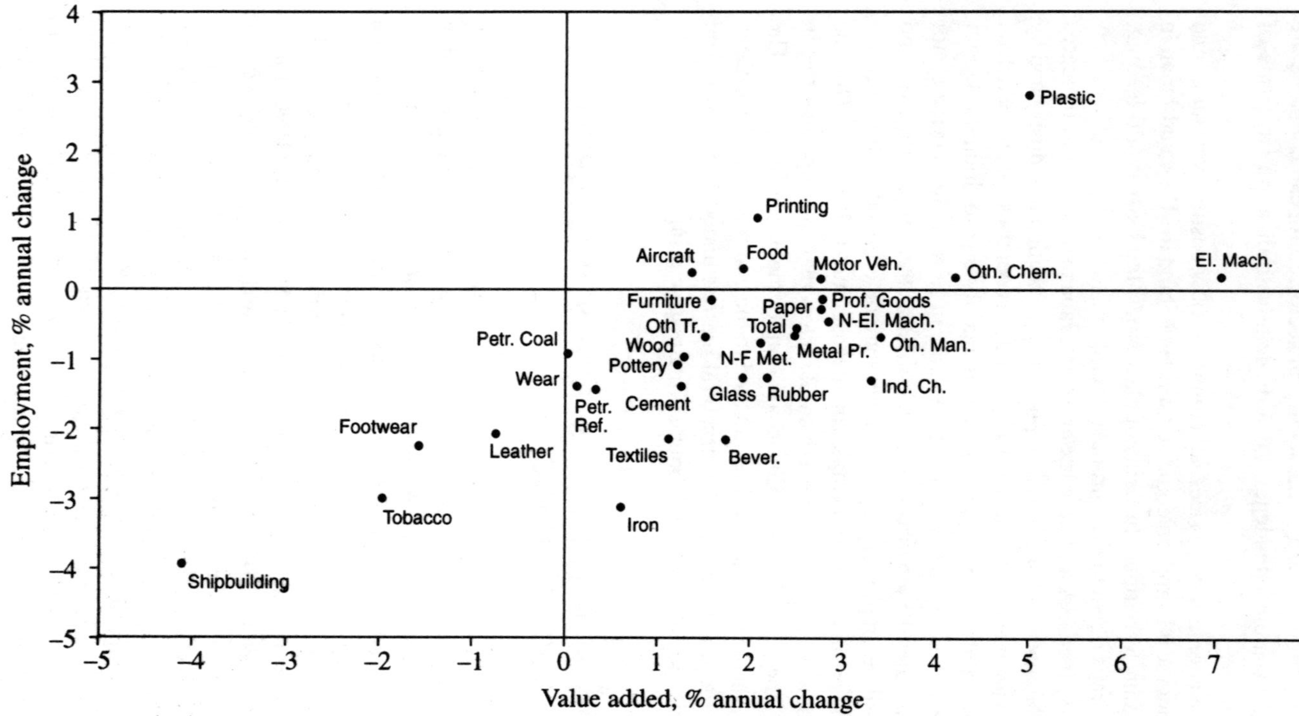

Note: Countries include the US, Japan, 13 EU countries and Norway (data are missing for Ireland and Luxembourg).

Figure 4.4 Changes in value added and employment in OECD countries, 1975–94

in the previous section. Across industries, a positive relation is evident, and the 31 sectors can be divided into the following three groups:

- *Growth sectors*, where value added and employment have both increased, including plastic, printing, other chemicals, electrical machinery, food, aircraft and motor vehicles.
- *Declining sectors*, with an opposite pattern of fall in both value added and employment, including leather, footwear, tobacco and shipbuilding.
- *Sectors in restructuring*, with growth in production and a fall in employment, including all remaining sectors and the values for total manufacturing.[3]

Looking at European countries, the dynamics of value added and employment show a rather different distribution, presented in Figure 4.5.[4] A systematic gap is evident between the overall OECD pattern and the much lower performances of Europe, which has most of its industries in the restructuring quadrant. Japan is the country with the highest number of sectors in the growth quadrant, while the USA has an intermediate position.

Focusing on the most recent years (1989–94), Europe's performance appears even more negative, with only one sector – plastic – left in the growth quadrant, while all the textile and leather sectors, all transport industries, non-electrical machinery, metal products, iron and steel, pottery and rubber move into the declining quadrant. Europe's distance from the USA and Japan has further increased in the 1990s.

The way the sectoral structure of national economies affects the aggregate employment outcomes shown above (the so-called *composition effect*) can be highlighted by the shares of the industries showing employment growth, moderate decline or strong decline in the 1975–94 period. Table 4.4 ranks the 31 industries according to their employment growth at the OECD level and shows their importance in the industrial structure (measured by the shares of manufacturing value added) of Europe, the USA and Japan in 1975 and 1994. The 'burden' on Europe's performance resulting from a larger presence in sectors with lower growth clearly emerges.

The industries with positive employment growth in the OECD group (as shown in Figure 4.4) in 1975 accounted for about 37–38 per cent of manufacturing value added in Japan and the USA, and just 32 per cent in Europe. At the end of the period all countries had increased their shares, with Japan at 49 per cent, the USA at 44 and Europe at 40 per cent. The stronger and increasing presence of the USA and Japan in the sectors with positive employment dynamics is a major factor behind their better aggregate employment performance highlighted above. In the industries with moderate employment decline no clear-cut picture emerges. Europe is strongly over-represented in

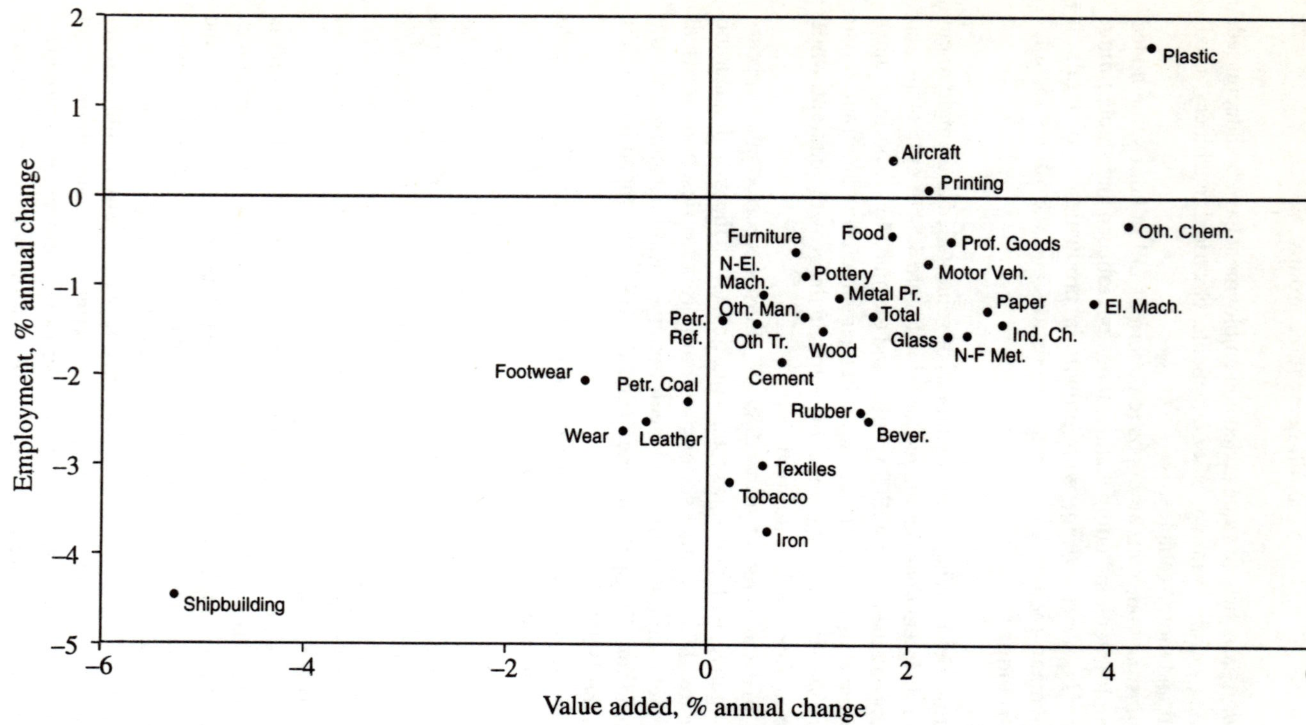

Note: Europe includes 13 EU countries and Norway (data are missing for Ireland and Luxembourg).

Figure 4.5 Changes in value added and employment in Europe, 1975–94

Table 4.4 Employment change and the sectoral structure of manufacturing industry

Sectors	Percentage rate of change of employment in the OECD, 1975–94	Percentage shares in total value added					
		Europe		United States		Japan	
		1975	1994	1975	1994	1975	1994
Sectors with employment growth							
Plastic products	2.84	1.9	3.1	1.4	3.3	3.8	3.0
Printing & publishing	1.06	4.0	4.4	7.8	6.1	5.3	4.4
Food	0.30	8.9	9.3	7.5	7.3	14.8	8.7
Aircraft	0.23	1.4	1.4	4.5	3.1	0.2	0.2
Electrical machinery	0.20	6.7	10.1	6.1	12.7	1.8	19.8
Other chemicals	0.19	3.3	5.3	4.8	6.2	4.8	5.0
Motor vehicles	0.15	6.0	6.7	6.5	5.3	6.1	8.3
Total		32.2	40.3	38.5	44.0	36.7	49.4
Sectors with moderate employment decline							
Professional goods	–0.15	1.5	1.7	4.4	4.0	0.9	1.5
Furnitures & fixtures	–0.15	2.3	2.0	1.5	1.6	2.1	1.2
Paper & products	–0.27	2.6	3.2	4.8	4.6	2.8	2.4
Non-electrical machinery	–0.45	11.5	9.4	8.3	11.5	7.6	10.6
Metal products	–0.65	8.3	7.8	6.6	7.1	5.2	6.1
Other transport	–0.67	0.5	0.4	0.2	0.1	0.6	0.6
Other manufacturing	–0.67	1.2	1.1	1.8	1.8	3.3	4.5
Non-ferrous metals	–0.76	1.3	1.5	2.1	1.6	2.0	1.5
Petroleum & coal products	–0.94	0.4	0.3	0.4	0.3	0.4	0.1
Wood products	–0.97	2.1	1.9	2.9	2.4	2.5	1.1
Pottery, china etc.	–1.08	0.8	0.7	0.1	0.1	0.4	0.3
Rubber products	–1.26	1.2	1.1	1.0	1.1	1.4	1.0
Glass & products	–1.26	1.0	1.1	0.9	0.6	1.1	0.8
Industrial chemicals	–1.29	4.4	5.6	3.9	5.2	5.6	4.1
Non-metallic products	–1.38	3.7	3.1	1.9	1.6	3.5	2.5
Wearing apparel	–1.41	3.2	2.0	2.1	1.9	3.3	1.3
Petroleum refineries	–1.46	3.8	2.9	2.3	1.6	1.2	0.6
Total		49.8	46.0	45.4	47.2	44.0	40.2
Sectors with strong employment decline							
Leather & products	–2.09	0.6	0.4	0.3	0.1	0.6	0.2
Beverages	–2.16	2.8	2.8	1.3	1.3	2.6	1.3
Textiles	–2.16	4.9	4.0	2.7	2.9	4.7	2.0
Footwear	–2.27	1.2	0.7	0.5	0.1	0.4	0.2
Tobacco	–3.03	2.0	1.5	5.0	1.3	0.6	0.3
Iron & steel	–3.15	4.7	3.9	5.4	2.7	8.3	6.0
Shipbuilding & repairing	–4.00	1.8	0.5	1.1	0.3	2.5	0.7
Total		18.0	13.7	16.1	8.8	19.6	10.6
Total manufacturing	–0.55	100	100	100	100	100	100

Notes:
Calculations from OECD STAN 1996 data.
Europe includes 13 EU countries (data are missing for Ireland and Luxembourg) and Norway.
Entries do not always sum up to total due to rounding.

the job-losing industries, with 14 per cent of its value added in 1994 (down from 18 per cent in 1975), against the 9–11 per cent of the USA and Japan. The two latter countries have been able to reduce much more rapidly their reliance on manufacturing activities with a marked job-destruction pattern. Thus, the combination of generally lower aggregate growth and higher shares of industrial employment in sectors dominated by labour saving seems to be at the root of Europe's unemployment problems. Structural factors, rather than labour market functioning, are a key explanation for the job losses in Europe's industry.

The second issue that deserves exploration is the direct impact of technological change at the sectoral level. A number of studies have been carried out on various aspects of the impact that technology has on employment performances.[5] In the rest of this section some recent results from these studies are presented.

Pianta, Evangelista and Perani (1996) have investigated the employment impact of technological and structural change in G-6 countries over the 1980s. Using sectoral data on patenting and gross investment as well as growth of value added, the study found that for the aggregate of G-6 economies in the 1980s, the sectors showing the highest rates of investment and innovation have experienced higher growth of output and employment. However, such a process becomes more uneven when the individual countries – especially the European ones – are considered, because the benefits of the compensation effects are distributed over countries as a result of the competitive process, which is affected by the sectoral specialization of individual countries. Countries with higher activity in the fields where new demand and employment grow faster are better positioned to capture the compensation effects, while economies dominated by declining industries are major losers in the employment effects of technological change. In Europe, and particularly in Italy, the 'virtuous circle' between technology, growth and employment is much weaker as investment and innovation have focused on the restructuring of traditional sectors, and are associated with large labour-saving effects.[6]

In Pianta (1996, 1998), the limitations of indirect indicators of technological change such as investment and patenting have been overcome with the use of a new data set derived from the European Community Innovation Survey (a standardized survey carried out in 1992). The employment impact of innovation across 20 manufacturing sectors has been investigated in Pianta (1996) for the cases of West Germany and Italy, testing the effects that the growth of value added and several indicators of innovative efforts have on changes in employment. The employment–value added relationship appears positive and significant only in the German case while innovation variables (such as R&D intensity and total innovation expenditure) negatively affect employment in both countries.

The most comprehensive results from this data set concern five European countries (Denmark, Germany, Italy, the Netherlands and Norway; all countries with adequate data from the Community Innovation Survey) and are presented in Table 4.5, based on Pianta (1998).

Employment changes in 1989–93 are explained by changes in demand and output (value added and exports), innovation intensity (total innovation expenditure per employee and some R&D variables), the prevalence of product innovations (share of R&D related to product innovations) and an output indicator such as the share of exports due to new products. When data for all countries are pooled together (see the first three columns of Table 4.5), value added, exports and product innovations have a positive impact on employment changes. Innovative expenditure per employee (or R&D expenditure per employee or per sale) and the share of new products in exports turn out with negative signs. Regressions for individual countries generally provide similar results, although not all variables are equally significant in all countries. Changes in demand, expressed either by value added or export, have a positive and significant impact in all countries except the Netherlands. Conversely, total innovation expenditure or R&D expenditure per employee has a significantly negative employment impact in Germany and Italy. The signs of the pooled results are confirmed in most country findings.

In explaining the changes (in most cases the decrease) of employment in European industries, demand factors and product innovations have a positive impact. However, a greater innovation (or R&D) expenditure is associated with worse employment outcomes, suggesting a dominant pattern of labour-replacing technological change. An interesting finding relates also to the export performance of European countries. The sectors where new products are more relevant in exports have worse employment outcomes, suggesting that in the industries that are more innovative and internationally open European competitiveness has been weak. On the other hand, employment losses have been limited in more traditional industries, with lower innovative intensity and technological content of exports. It appears from these results that Europe relies relatively little on technological advantages – launching new products and opening new international markets – but rather on other factors, such as the defence of market shares with cost reductions, international relocation of production and process innovations. In the longer term such competitive strategies can hardly support a stable base of industrial employment.

The evidence emerging from these relationships is that, in the cycle of growth which includes the recession of the early 1990s, demand factors[7] have been crucial in sustaining employment, while technological factors show contrasting effects for product innovations (positive) and general technological efforts (negative).

Table 4.5 The employment impact of economic and technological change

	Pool of 5 countries			Denmark	Germany		Italy			Netherlands	Norway
	a	b	c		a	b	a	b	c		
Constant	–0.022 (–2.27) **	–0.019 (–2.453) **	–0.022 (–2.510) **	–0.058 (–2.945) ***	0.023 (1.976) *	0.031 (3.320) ***	–0.004 (–0.790)	–0.005 (–0.698)	–0.005 (–0.669)	–0.018 (–0.209)	–0.027 (–1.925) *
Change of value added	0.285 (3.88) ***	0.28 (3.813) ***	0.278 (3.785) ***	0.25 (2.064) *	0.604 (3 712) ***	0.715 (4.993) ***	0.158 (1.647)			0.349 (0.258)	
Change of export	0.135 (3.903) ***	0.139 (4.010) ***	0.146 (4.144) ***						0.066 (1.952) *		0.218 (3.678) ***
Total innov. exp. per employee	–4.08E-07 (–2.146) **						–1.31E-06 (–3.600) ***	–1.14E-06 (–2.975) ***		–2.52E-07 (–0.394)	
Share of innovative export	–0.021 (–1.757) *	–0.022 (–1.725) *	–0.021 (–1.687) *	–0.035 (–2.339) **				–0.034 (–1.882) *	–0.051 (–2.687) **		–0.069 (–2.504) **
Share of R&D related to product innovation	0.027 (1.867) *	0.027 (1.875) *	0.028 (1.921) *	0.103 (3.460) ***						0.074 (0.572)	0.039 (1.536)
R&D expend. per employee		–4.00E-07 (–2.001) **				–8.580E-0.6 (–3.428) ***					
R&D expend. on sales			–5.50E-02 (–1.945) *		–7.32E-01 (–1.951) *						
Adjusted R-squared	0.27	0.26	0.26	0.5	0.49	0.63	0.51	0.53	0.42	0.41	0.53
F	6.77 ***	6.61 ***	6.55 ***	5.06 **	8.74 ***	15.2 ***	9.39 ***	10.13 ***	6.54 ***	0.22	5.68 ***
Number of cases	81	81	81	19	21	21	21	21	21	19	19

Notes:
Dependent variable: rate of change of employment (1989–93), 21 manufacturing industries.
t-statistics shown in brackets. Significance levels: *90%; **95%; ***99%.
a, b and c refer to different specifications of the model under test.

Source: Pianta (1998).

A generalization for European countries, the USA and Japan has also been carried out on the basis of the relevance of industries dominated by the employment-friendly model of high demand growth and product-oriented innovations, or by the labour-displacing pattern of low demand and process innovations. Again, European countries emerge with the worst positions both in terms of their industrial structure (with a large weight of the latter group) and in terms of their relative growth performances (with slower growth of value added and employment in the former group) (see Pianta, 1998).

These findings and the surprisingly widespread negative links between the technological intensity of industries and their employment performance in Europe suggest that more attention has to be paid to the specific nature of technological change, with a major distinction between product and process innovations. This issue, first raised in the study of Italy in the 1980s (Vivarelli, Evangelista and Pianta, 1996), now appears to characterize the bulk of European innovation performances, suggesting that a major reconsideration of the relevance of technological unemployment is now needed.

5 CONCLUSIONS

The first result of this study is that Japan and the USA on the one hand and European countries on the other deeply differ in their capacity of job creation both in absolute terms and in terms of employment sensitivity to economic growth. Differences in the employment outcomes of production can be related to the different industrial structures across OECD countries. Indeed, European specialization tends to be concentrated in sectors where labour-saving process innovations are dominant, while European countries reveal structural weaknesses in the sectors where product innovation is prevailing.

On the whole, the new division of labour induced by European integration and international competition seems to relegate Europe to a position where industrial output growth is less and less linked to employment growth. The meaning of 'growth' itself is changing, losing much of its employment and 'social' content, and this should be taken into account when assessing the impact of issues such as competitiveness and specialization. From this evidence we can draw some lessons on the impact of innovation and competitiveness, on the importance of demand and on the need for coherence between structural (and industry-specific) policies and macroeconomic policies.

First, the performance of Europe's industries shown in this chapter clearly suggests that industrial and competitiveness policies across European countries should take into account the employment dimension in trying to shape Europe's industrial future. In spite of the unpopularity in recent years of selective industrial policies, some 'targeting' of sectors with high growth and employment potential, high externalities and high capacity for learning and for developing and adopting innovations oriented towards new products is required. This is particularly important in ICT-related sectors, where the aim should be the emergence of new economic activities, able to meet economic and social needs, to develop new markets and offer new growth and employment opportunities (see Chapter 5 in this volume).

Second, focusing the analysis and policy on the supply structure alone leads to a one-sided and inadequate view of the complex links between innovation and structural change and employment. We cannot expect Say's law to operate when a new technological paradigm is emerging and deep changes in the economic structure are under way. In spite of the efforts on the supply side to increase competitiveness, exports growth is likely to be unable, by itself, to sustain aggregate demand, which remains a crucial field for national and European growth and employment policies. In recent years the continuing anti-inflationary stance of monetary and fiscal policy has led to serious deflationary effects in Europe, reducing demand, depressing growth and increasing unemployment. The lack of demand has been particularly serious in the industries growing faster at the world level, and particularly in industries related to the new information and communication technologies. A growth strategy, with an active demand policy, is required in Europe also for implementing the structural and technological changes discussed at length by the other chapters of this book, as demand-side aspects have appeared to be crucial for the successful performance of industrial sectors.

Third, the coherence between the objectives of industrial and innovation policy and those of aggregate economic growth is essential. Without an active industrial and innovation policy, the evolution of Europe's economic structure is likely to lead to a further weakening of competitiveness and specialization patterns. Without a more expansionary policy, the impact of supply-side measures is likely to emphasize the losses of old activities and specialization and the cost-cutting effects of actions for competitiveness, rather than the gains in new activities, competences and employment.

This challenge is all the more important with the emergence of the European economic and monetary union, which represents a great risk and a great opportunity. The risk is that the fuller integration provided by the single currency will lead to a continuation of deflationary policies and to an even faster process of restructuring and concentration of European industry, with continuing losses in the industrial base and little growth in services. One estimate puts at two million jobs the employment losses that might result from the monetary union by the year 2000 (*Business Week*, 1998, p. 43). The opportunity, on the other hand, is for the creation of a new range of industrial strategies and demand policies able to reorient growth towards much needed emerging and employment-intensive new activities.

NOTES

1. Padalino and Vivarelli (1997) for instance, found a negative working time trend in West Germany and France over the period 1960–94. Problems in data availability do not allow us to investigate these aspects in this chapter.
2. The OECD data on employment growth in industry and services further support this argument. In the European Union (including 13 countries) the number of jobs in industry fell at average annual rates close to 1 per cent in the 1970s and 1980s, and at above 3 per cent in the 1990–94 period. In the USA and Canada the 1970s saw a growth above 1 per cent, the 1980s a stability, and the 1990s a 1 per cent decrease. In Japan, employment grew steadily by 0.7–0.8 per cent a year from 1970 up to 1994. In services, Europe also had an employment performance worse than its competitors, with increases close to 2 per cent a year in the 1970s and 1980s and at 0.6 per cent in the 1990s. In North America, service jobs grew at above 3 per cent in the 1970s and at 1.7 per cent in the 1990s, with a pattern slightly higher than in Japan (see OECD, 1996A, p. 12).
3. The full list of sectors in restructuring is the following: furniture, non-electrical machinery, paper, rubber, other transport, other manufacturing, professional goods, metal products, non-ferrous metals, pottery, glass, wood products, industrial chemicals, cement, textiles, wearing apparel, petroleum and coal, petroleum refineries, other transport equipment, beverages and iron and steel.
4. For total manufacturing, the rates of change of value added and employment differ significantly across the three areas. Employment has increased by an annual rate of 0.43 per cent in Japan and 0.11 per cent in the USA, and has fallen by –1.32 per cent in Europe. Value added has increased by 3.86 per cent in Japan, 2.94 per cent in the USA, and 1.62 per cent in Europe.
5. The results emerging from the specific work on employment within this research project are presented in Petit and Soete (1999).
6. The regression results show that the growth of value added appears moderately significant only in the USA and Germany. Investment has the greatest and more generalized positive impact on employment growth, with the only exceptions being France and Italy. Patents show a positive and significant link to employment only in Germany and France. The lack of association for Japan may be due to the high growth rates of patenting in the USA experienced by most sectors, even of lower technological intensity. Overall, the regressions largely account for the employment changes of all countries, with the exception of Italy where no variable is significant.
7. On the demand side, it has been pointed out that the structure of demand, price elasticities and income elasticities are important for industry-specific employment trends. The dynamics of productivity also plays a key role, and it has been shown that the positive correlation between industry-specific productivity growth and employment has turned into a negative one in more recent years (Appelbaum and Schettkat, 1995).

BIBLIOGRAPHY

Amable, B., R. Barré and R. Boyer (1997), *Les systèmes d'innovation à l'ère de la globalisation*, Paris: Economica.

Appelbaum, E. and R. Schettkat (1995), 'Employment and Productivity in Industrialized Economies', *International Labour Review*, **134** (4–5), 605–23.

Berman, E., J. Bound and Z. Griliches (1994), 'Changes in the Demand for Skilled Labor within U.S. Manufacturing: Evidence from the Annual Survey of Manufactures', *Quarterly Journal of Economics*, **109**, 367–98.

Betts, J. (1997), 'The Skill Bias of Technological Change in Canadian Manufacturing Industries', *Review of Economics and Statistics*, **79**, 146–50.

Business Week (1998), 'The Euro. Will it Create a New European Economy?' Special issue, 27 April.

Centre for Economic Policy Research (1995), *European Unemployment: Is There a Solution?*, London: CEPR.

Doms, M., T. Dunne and K. Trotske (1997), 'Workers, Wages, and Technology', *Quarterly Journal of Economics*, **112**, 253–89.

Dosi, G. (1988), 'Source, Procedures and Microeconomic Effects of Innovation', *Journal of Economic Literature*, **26**, 1120–71.

Drèze, J.H and C.R. Bean (eds) (1990), *Europe's Unemployment Problem*, Cambridge, MA: MIT press.

Fagerberg, J. (1994), 'Technology and International Differences in Growth Rates', *Journal of Economic Literature*, **32**, 1147–75.

Freeman, C., J. Clark and L. Soete (1982), *Unemployment and Technical Innovation*, London: Pinter.

Freeman, C. and L. Soete (eds) (1987), *Technical Change and Full Employment*, Oxford: Basil Blackwell.

Freeman, C. and L. Soete (1994), *Work for All or Mass Unemployment? Computerised Technical Change into the Twenty-first Century*, London: Pinter.

Katsoulacos, Y.S. (1984), 'Product Innovation and Employment', *European Economic Review*, **26**, 83–108.

Katsoulacos, Y.S. (1986), *The Employment Effect of Technical Change*, Brighton: Wheatsheaf.

Layard, R., S. Nickell and R. Jackman (1991), *Unemployment: Macroeconomic Performance and the Labour Market*, Oxford: Oxford University Press.

Marx, K. (1961), *Capital*, Moscow: Foreign Languages Publishing House (first edn 1867).

Marx, K. (1969), *Theories of Surplus Value*, London: Lawrence & Wishart (first edn 1905).

OECD (1994), *OECD Job Study. Facts, Analysis and Strategies*, Paris: OECD.

OECD (1996a), *Technology, Productivity and Job Creation, Highlights*, Paris: OECD.

OECD (1996b), *Technology, Productivity and Job Creation, Analytical Report*, Paris: OECD.

Padalino, S. and M. Vivarelli (1997), 'The Employment Intensity of Economic Growth in the G-7 Countries', *International Labour Review*, **136**, 191–213.

Pasinetti, L. (1981), *Structural Change and Economic Growth*, Cambridge: Cambridge University Press.

Petit, P. (1995), 'Employment and Technological Change', in P. Stoneman (ed.), *Handbook of the Economics of Innovation and Technological Change*, Amsterdam: North Holland, pp. 366–408.

Petit, P. and L. Soete (eds) (1999), *Technology and the Future of European Employment*, London: Routledge.

Pianta, M. (1996), 'Technology and Jobs in the 1990s. The Results of the German and Italian Innovation Surveys', paper presented at the conference 'Technology, Economic Integration and Social Cohesion', CEPREMAP, 22–23 November, Paris.

Pianta, M. (1998), 'Innovation, Demand and Employment', paper presented at the conference 'Technology, Economic Integration and Social Cohesion', CEPREMAP, 15–16 May, Paris.

Pianta, M., R. Evangelista and G. Perani (1996), 'The Dynamics of Innovation and Employment: An International Comparison', *Science Technology Industry Review*, Paris: OECD, pp. 67–93.

Rosenberg, N. (1976), *Perspectives on Technology*, Cambridge: Cambridge University Press.
Stoneman, P. (1983), *The Economic Analysis of Technological Change*, Oxford: Oxford University Press.
Sylos Labini, P. (1969), *Oligopoly and Technical Progress*, Cambridge, MA: Harvard University Press (first edn 1956).
Vivarelli, M. (1995), *The Economics of Technology and Employment: Theory and Empirical Evidence*, Aldershot: Edward Elgar.
Vivarelli, M., R. Evangelista and M. Pianta (1996), 'Innovation and Employment in Italian Manufacturing Industry', *Research Policy*, **25**, 1013–26.

5. Europe and the information and communication technologies revolution

Bent Dalum, Chris Freeman, Roberto Simonetti, Nick von Tunzelmann and Bart Verspagen*

1 INTRODUCTION

Historians and economists have often been reluctant to use the expression 'revolution' to describe widespread and relatively rapid structural changes in the economy. Even in the case of the 'Industrial Revolution' revisionist historians have been anxious to downplay both the speed and the scale of the transition. They point out that many changes took place already in the early part of the eighteenth century and that the acceleration of economic growth, which was at one time widely believed to have taken place in Britain in the 1780s and 1790s, was not quite so rapid as was once estimated. Furthermore, they insist that only a few industries made the transition to factory production before the middle of the nineteenth century and that a large part of the population continued to work in agriculture. Nevertheless, most historians have followed Landes (1969) and Mokyr (1990) in their continued adherence to the expression 'industrial revolution' as the most succinct, convenient and accurate summary of the manifold social and technical changes which took place in many European countries and in the United States in the nineteenth century.

At one time also, there was some resistance to the use of expressions such as the 'information revolution' or the 'computer revolution' to describe the contemporary process of structural change affecting many countries. Now that computers have become commonplace in almost every workplace, classroom and office, as well as many homes, there is much less resistance to the use of these expressions which have indeed entered everyday language. The 'millennium bug' has brought home to almost everyone the dependence of our industries and services, as well as both civil and military government activities, on efficient computer software.

Previous technological revolutions such as electrification or mechanization have also affected numerous industries and services but none has affected every *function* within each industry in the way that computers do. Most research, development and design work has now become dependent on the use of computers. So too have many production processes and much market research and marketing organization, as well as numerous routine administrative and accounting activities. The convergence of telecommunications and entertainment services with computer technology is apparent in almost every household, as well as the largest multinational companies.

Yet, while the scale and range of computerization are no longer a matter of dispute, controversy still surrounds the economic impact of this information revolution, notably the productivity gains which might have been expected from such a widespread process of technical and economic change. This problem has become widely known as the 'Solow Paradox' since that famous economist pointed to the apparent contrast between the universal availability of computers in the United States and the very limited evidence of productivity gains from their use. Thus, while the diffusion of computers (and related technologies) throughout the production process and/or daily life is inevitable, it is not so clear whether or not we can speak of a 'revolution' in terms of its economic impact. In the last two decades there have actually been several industries that *have* exhibited extremely high rates of productivity growth within a sea of relatively sluggish performers. Such gains as have been achieved in overall productivity growth in the economy have been largely due to a few star performers, notably the computer industry itself, the semiconductor industry, electronic consumer goods and telecommunications, and to the increasing weight of these sectors in the economy as a whole.

Looking back to previous technological revolutions, a similar pattern can be observed. A cluster of new very fast-growing industries burst into economic life and was characterized both by very high productivity growth and by high aggregate growth. Schumpeter (1939) characterized these periods of explosive growth of new leading industrial sectors as 'successive industrial revolutions' and associated the rise to dominance of the new clusters with the long cycles or 'Kondratieff waves' identified by many economists in the late nineteenth and early twentieth centuries. In these terms the 'Computer Revolution' is the latest in a series of industrial revolutions.

It is increasingly clear from work in a 'neo-Schumpeterian' tradition, that the economic impact of such 'revolutions' depends on a profound process of institutional and organizational change throughout the social system, including both the more 'traditional' areas, and the leading fast growth industries. In other words, whether or not we will 'see computers in the productivity statistics' depends on how successful other sectors will be in adopting and implementing the new technology. This involves a host of factors which economists usually refer to as 'institutions', for instance, education of the labour force, internal and external organization of firms and management practices, (de)regulation of certain sectors such as telecommunications, etc. It was primarily Carlota Perez (for instance, Perez, 1983) with her theory of changes in 'techno-economic paradigm' who pioneered this wider conception of the importance of social and institutional change in the emergence and diffusion of major new technologies. The diffusion of a major new technology system is a gigantic social learning process and is necessarily accompanied by all the delays and lags which attend any learning process on this scale. These problems are most apparent in the diffusion of the technology beyond the initial fast growth sectors to other industries and services.

We will argue that the ICT revolution is a real phenomenon, and that Europe's future well-being will to a large extent depend on its ability to develop and exploit the opportunities inherent in ICT. The extent to which Europe will succeed in this depends on many factors that we cannot deal with in a fully satisfactory way in this chapter. What we hope to do, however, is to prepare the ground for such a discussion by providing a broad comparative analysis of production and use of ICT in the developed world.

The focus on the economic and technological competitiveness of the European ICT sector is in line with our overall quantitative economic perspective in this volume. At the same time, however, it is clear that the trends of structural change and diffusion of a new techno-economic paradigm are very hard to capture in the statistics. The pervasive nature of ICT makes it hard to isolate its impact from the 'more traditional' parts of the economy. Also, existing statistical classifications cannot cope too well with new activities, especially if they emerge in very different sectors of manufacturing and services.

The analysis starts, in section 2, with a broad overview of the development of ICT as a technology. This section outlines how ICT evolved in response to the needs of different kinds of companies, and convergence of different technologies (semi-conductors, automatic data processing, telecommunications) took place. It also describes how, in the process of diffusion of ICT, virtually all functions within the firm became affected by ICT (as touched upon above already). Section 3 then presents a short overview of some quantitative trends in technology, international trade and production, which clearly establish the claims made in this introduction, as well as section 2, on the overall importance of ICT in the economy. European competitiveness in ICT is the topic of section 4. Here we review the available evidence on European performance in technology, trade and production. We consider both the hardware (manufacturing) parts of ICT, and the services content. Section 5 concludes the argument, and draws some preliminary conclusions on European policy in this area.

2 MICROELECTRONICS AND THE EMERGENCE OF NEW ICT

Information technology in the form of computerization generally entered firms first through the function we describe as 'administration' – essentially to facilitate 'back-office' operations such as dealing with payrolls. Computers represented a considerable advance in office machinery, but at the same time could by and large be integrated into existing administrative structures. This was an important reason why office machinery firms, and especially IBM, came to dominate the early growth of the computer equipment industry. The emphasis fell on mainframe computers, which rapidly increased in processing power but, despite an equally rapid fall in cost per bit of information processed, remained very expensive items of equipment to purchase outright. The threshold scale of output at which it became economic to purchase even a moderate-sized computer remained very high, and hence static barriers to entry in terms of acquiring such machines remained high. In the 1960s, efforts were made to decrease such outlays through the introduction of 'time-sharing', which allowed a particular machine to have multiple users and thus brought down the barriers to entry. This kind of development also facilitated the rise of the software industry, partly established to operate effective time-sharing schemes.

Developments in hardware also continued to reduce barriers to entry. The introduction of minicomputers by Digital Equipment and others in the mid-1960s reduced the initial purchase price, as, much more dramatically, did the arrival of the personal computer (PC) in the late 1970s. With the latter, the purchase price had more or less ceased to be an entry barrier for firms to acquire computers. These developments were of course underpinned by the exceptionally rapid innovation in the semiconductor industry, first with the transistor, then with the integrated circuit (IC), and finally in the early 1970s by the microprocessor (MPU).

This continual decrease of the barriers to entry of using computer technology led to at least three important trends. First, computer technology entered virtually all parts of firms' operations. Second, partly as a result of the first, specialization led to a trend of outsourcing computer- and ICT-related services. Third, the widespread use of computers and electronics opened up possibilities for technological and economic convergence between activities that had hitherto largely been separated, that is, telecommunications, electronics and publishing.

Regarding the first of these trends, one can note several important aspects of firms' activities in which ICT plays an increasingly important role. In the process of generation of new technology, these technologies are increasingly used to reduce the cost of trial and error search through the use of simulation on high-speed or supercomputers of complex physical, chemical and biological interactions. New instruments based on ICT have also considerably improved routine technological operations such as quality testing. In the production process, ICT has substantially changed the organization of the shop-floor in most manufacturing industries. The development of 'mechatronics', that is, fusing mechanical engineering and electronics, led to electronics and software being 'embedded' into tools and equipment, resulting in speeding up existing production runs. The ability to programme and reprogramme computer-controlled equipment allows firms to reap economies of scope through the reduction of downtime in changing lines of production (shifting to new machining standards or introducing new components), improvement of the coordination of single machines or, with networking, of whole systems. 'Vertical product differentiation' in the form of varying qualities being produced in response to demand shifts can therefore be much accelerated. In practice, the take-up of such 'flexible automation' in various forms – both stand-alone as in specific CNC tools and networked as in computer-integrated manufacturing (CIM) – seems to have been considerably faster in large firms as compared to SMEs in most countries and industries.

With regard to the second trend, outsourcing, the importance of the role of ICT within firms has generated the need for competent and cost-effective development and implementation of IT systems in firms, which could not always be provided internally. Outsourcing of particular functions and services, therefore, has become very extensive in the 1990s. First, this concerns services directly related to computer technology, such as maintenance, and tailored software development. The specialist computer consultancy firm Gartner Group now estimates that support costs – as opposed to purchase costs – amount to about three-quarters of the lifetime costs of a PC, and a significant portion of this is outsourced to specialized companies.

However, ICT also enables outsourcing beyond this direct form. For example, routine testing has long been quite suited for 'outsourcing' to independent laboratories in which static R&D economies are attainable (Mowery and Rosenberg, 1989), and this now seems to be spreading again, under the influence of networking through ICT. Thus for a particular product (say bread) it may pay to outsource routine testing to laboratories dedicated to testing a variety of products (foodstuffs) rather than to interrupt production in the firm responsible for producing the wheat in order to carry out sporadic testing. By concentrating the testing elsewhere, dynamic scale economies may be obtained in real time (time-saving) and in historical time (learning effects) in the laboratories, while the firm does likewise in actually producing bread.

With regard to the third trend, convergence, this appeared first between the fields of computer technology and communications. The appearance of a PC on every desk in the company led to the necessity to link these machines together in more effective ways. From the 1960s, mainframe computers were being linked by means of the telephone and modems, but this was a laborious and lengthy procedure. Opportunities for time-saving advances, however, also lay in improvements to the communications technologies themselves. At a somewhat more sedate pace, telecommunications was shifting from electro-mechanical switching to semi-electronic and then fully electronic systems, and from low-capacity wired transmission to high-capacity alternatives such as satellites and fibre optics. In the 1980s there was a rush of mergers and acquisitions by both IT and telecomms companies looking for 'convergence' from the common factor of digital electronics. However, this 'technological convergence' did not imply convergence in the business field to the same extent. Scope economies in producing convergence turned out to be rather limited (Soete and von Tunzelmann, 1987; Duysters, 1996). First, there was little convergence in markets – telecomms companies produced switches mainly for large customers such as PTTs or the military, with large batches and high customization. Second, there was less convergence in technologies than had been expected. Telecomms companies pursued strategies in which the communications network would remain the central focus, which steered them towards an 'information super-highway' based on ISDN (Integrated Services Digital Network). IT companies instead put the main focus on 'intelligent' terminals, and sought to link these through Local Area Networks (LANs), Wide Area Networks (WANs) and the like. With this continuing divergence, most of the alliances and acquisitions undertaken with convergence in mind soon fell apart.

By the 1990s, a new area of convergence in products was emerging, now also incorporating the field of publishing and entertainment. At first, this was mainly the outcome of new entrants focusing on areas such as multimedia, and filling a gap somewhere between specialized information and communication products. The objective in multimedia was to be able to transmit not just data but voice and images (both still and moving) at high speed between terminals. The communications side remained something of an obstacle to all attempts to compress such data (especially moving images) but the contemporaneous development of the Internet – again outside of the traditional carriers – appeared to provide a solution. The exceptional growth of the Internet in the mid-1990s is too well known to require further comment.

In terms of our analysis, these three trends have several important implications. First, the increasingly widespread use of ICT implies that the product markets, both in manufacturing and services, have grown at very high rates. This implies important business opportunities for those firms (or countries) specialized in the 'right markets'. Second, it leads to an increasing role for services in the economy, particularly ICT services themselves, but also industries such as finance and insurance, and media, which to a large extent rely on ICT for their production process. Third, related to the convergence of different technological fields, firms, and at a higher level of aggregation, countries, originating in different parts of the ICT field are in different positions regarding the key development. For example, and as will become more clear in the next sections, countries specialized in telecommunications can be shown to have quite different performance in ICT than countries specialized in computers. The next two sections will pick up these three points by relating them to the actual data on trends in the world economy, specifically the role of Europe in it.

3 ICT AND THE WORLD ECONOMY: SOME TRENDS

This section looks at the growth of ICT and the markets for ICT products. Our primary indicator of technological developments will consist of patenting activity in the USA. Basically, patents are related to intellectual property, which is only one aspect of the innovation process in general. Nevertheless, patenting in the USA, which constitutes the largest national market in the world over the postwar period, may reveal some useful information, both on the technological specialization of countries (see section 4), and trends in technological opportunity in general.

Table 5.1 shows that many of the ICT-related patent classes are among the ones with most rapid growth of the number of patents in the last 20 years, particularly so in the 1980s.[1] There is a clear difference in the fast-growing classes between the 1970s and the 1980s, and the difference suggests that there has been a break with the past. In the 1970s, many of the fast-growing classes were related to traditional technologies (for example, internal combustion engines; see Meliciani and Simonetti, 1998), while in the 1980s ICTs dominate the fast-growing patent classes. We interpret this as an indication of high technological opportunities in the ICT area. It indicates that research in this area is growing, that is, that firms judge the development of new ICT systems to be profitable. This confirms the technological dynamism that is associated with the idea of ICT as a technological revolution.

Table 5.2 presents a similar exercise for international trade data (value of exports from OECD countries). It presents the 20 product classes with most rapid growth between 1961 and 1994. Many of the classes appearing in this list have an obvious ICT-related content. The two classes at the top, computers and semiconductors, are core classes with ICT-related equipment, as is telecommunications equipment, which appears at rank 8. Many of the other high growth classes have a clear ICT content, such as electromedical equipment, measuring and control equipment, and other electrical equipment.

Table 5.1 Rates of growth and rank of ICT patent classes in periods 1969–74 to 1979–84 and 1979–84 to 1989–94

Code	Class	Growth 1969–74 to 1989–94 (%)	Rank	Growth 1969–74 to 1979–84 (%)	Rank	Growth 1979–84 to 1989–94 (%)	Rank
82	Semiconductors	165	4	–14	38	209	3
85	Data processing systems	155	5	–14	39	197	4
77	Television, facsimile	140	7	–2	21	145	7
72	Laser technology	136	8	–31	67	243	2
73	Telecommunications eqp	133	10	–2	20	137	8
78	Acoustics	98	14	14	10	73	23
83	El. & electronic instruments	51	22	–25	57	102	11
79	Electrical devices & systems	31	33	–29	62	84	17
74	Radio comm. systems	21	40	–33	70	80	19
76	Machine controls	0	58	–42	86	73	24

Source: SPRU-OTAF database.

Table 5.2 Growth rankings and export shares for OECD high-growth products 1961–94

High-growth rank 1961–94	Rank 1961–73	Rank 1973–94	OECD export shares 1961	1973	1984	1994
1 Semiconductors	3	3	0.3	0.7	1.5	2.6
2 Computers	12	1	0.6	0.9	3.1	4.1
3 Orthop. & hearing aids	10	2	0.0	0.0	0.1	0.2
4 Electromedicals	6	5	0.1	0.2	0.3	0.4
5 Furniture	2	8	0.3	0.6	0.7	1.0
6 Non-elec. medical eq.	11	4	0.1	0.2	0.3	0.5
7 Plastic materials	5	15	1.0	1.8	2.1	2.5
8 Telecommunications	15	6	0.9	1.2	1.5	2.2
9 Measur. & control eq.	13	10	0.4	0.5	0.9	0.8
10 Motor vehicles	8	18	6.1	9.7	11.4	12.3
11 Organic chemicals	9	17	1.3	2.0	2.7	2.6
12 Mach. prod. & dist. el.	25	9	1.7	1.8	2.1	2.9
13 Other electrical eq.	23	11	1.0	1.2	1.3	1.8
14 Pharmaceuticals	32	7	1.1	1.1	1.1	1.9
15 Pumps & centrifuges	21	16	0.8	1.0	1.0	1 3
16 Clothing	19	21	1.3	1.8	1.6	1.9
17 Photo, optic. & watch	22	20	1.8	2.2	2.3	2.6

Table 5.2 continued

High-growth rank 1961–94	Rank 1961–73	Rank 1973–94	OECD export structure 1961	1973	1984	1994
18 Other non-elect eq.	16	24	1.5	2.0	1.7	2.2
19 Power generators	24	19	1.9	2.1	2.4	2.6
20 Other manufacturing	34	12	5.1	4.7	5.4	6.7

Source: The IKE Trade Database. The OECD export structure is calculated as the distribution of exports from OECD countries (intra-EU trade included).

International trade data, however, capture only a limited part of total world production value in ICT industries. An important part of ICT sales stems from domestic sources, and the services part of ICT is not adequately captured in the statistics on trade. Table 5.3 provides a more comprehensive picture of world markets for ICT in the OECD countries. The table illustrates some of the trends identified in the previous section, such as the shift away from large computers (multi-user systems) towards stand-alone PCs and workstations, the increasing importance of software and services, and the increasing role of data communication equipment.

The last issue that will be covered here relates to the use of ICT. The data in this area are scarce, but Table 5.4 lists some of the available indicators. These generally point in the same direction. The most striking feature is perhaps that the diffusion of ICT-related products is very high in the Nordic countries, and very low in the southern European countries. This result is found for each of the five indicators. The larger European countries show a rather 'average' performance. Germany ranks high with regard to ISDN, but low in mobile communications and the number of Internet hosts. In France, the use of Internet is very low. Some of the smaller but highly developed European countries, such as Switzerland and the Netherlands, also rank high. This indicates that in the European situation, there is a clear correlation of the spread of ICT with the level of economic development. As will be argued in more detail below, this pattern may have implications for the supply side of European ICT.

Overall, Europe is clearly lagging behind relative to the USA, except with regard to the spread of ISDN lines. Compared with Japan, Europe's position is more favourable with regard to the use of personal computers and ISDN, but not so for the other indicators. However, as mentioned, the European position results from rather diverging trends at the country level. In some European countries, ICT has reached very high levels of diffusion, and this points to a relatively advanced use of the new technologies. Especially as compared to Japan, the more advanced European countries rank high.

Table 5.3 ICT markets in selected OECD countries, 1987 and 1995, by segment. Levels in current million US$

		USA	Japan	Europe	OECD-24
ICT market size					
	1987	105 255	35 794	68 064	220 261
	1995	212 745	82 296	136 373	460 305
	Growth	*9.2*	*11.0*	*9.1*	*9.7*
Shares of segments (%)					
Multi-user systems	1987	23.6	40.0	28.1	27.9
	1995	9.6	22.8	12.9	13.0
	Growth	*–2.4*	*3.4*	*–1.0*	*–0.3*
Data comms equip.	1987	3.5	2.4	1.7	2.8
	1995	4.5	3.2	3.9	4.2
	Growth	*12.7*	*15.0*	*20.8*	*15.4*
PCs & workstations	1987	23.4	12.9	21.0	21.1
	1995	30.4	27.2	26.0	28.7
	Growth	*12.8*	*21.8*	*12.0*	*13.9*
Total hardware	1987	50.4	55.3	50.8	51.8
	1995	44.5	53.2	42.8	45.9
	Growth	*7.5*	*10.4*	*6.8*	*8.0*
Software	1987	15.8	8.2	15.1	14.1
	1995	20.5	11.1	22.3	19.1
	Growth	*12.8*	*15.2*	*14.5*	*13.9*
Services	1987	33.8	36.5	34.1	34.1
	1995	35.0	35.8	34.9	35.0
	Growth	*9.7*	*10.7*	*9.4*	*10.0*

Notes:
Europe is EU-14 (excl. Luxembourg) plus Norway and Switzerland.
Growth rates are annual percentages of market size; values for 1995 are estimates.

Source: OECD Information Technology Observatory.

The trends outlined in this section confirm the importance and rapid growth of ICT as stressed in the introduction. They also confirm the emphasis in ICT markets shifting towards decentralized systems and a convergence of different technological and business fields. At the same time, they underline, however, that the spread of ICT is by no means a homogeneous trend that takes place in all parts of the world at the same pace. Particularly in Europe, there are huge differences with regard to the spread of the use of ICT. Northern Europe is taking a (world) leading role in this respect, but southern Europe is clearly lagging behind.

Table 5.4 Some indicators on the use of ICT, 1996

	Cellular phone (1)	Internet hosts (2)	Internet users (2)	Personal computers (1)	ISDN subscribers (2)	Overall rank (3)
Norway	28.7	34.2	113.8	28.5	10.0	1
Germany	7.1	8.4	30.5	23.3	23.7	9
Finland	29.2	61.3	167.8	19.5	5.7	2
Switzerland	9.3	18.7	52.1	40.9	17.7	5
Denmark	25.0	20.3	57.0	30.4	5.7	3
UK	12.2	12.4	43.0	19.3	4.5	10
Netherlands	5.2	17.4	58.0	23.2	6.4	8
Sweden	28.2	26.9	90.5	21.5	2.2	6
France	4.2	4.1	8.6	15.1	7.3	15
Austria	7.4	11.0	37.2	14.9	5.2	12
Belgium	4.7	6.4	29.5	16.7	5.4	13
Italy	11.2	2.6	10.2	9.2	1.8	16
Spain	3.3	2.9	13.4	9.4	0.9	18
Portugal	6.7	2.4	23.2	6.7	2.0	17
Ireland	8.2	7.6	22.7	17.0	0.0	14
Greece	5.3	1.6	14.3	3.5	0.0	19
Europe (4)	8.8	8.7	29.7	16.9	8.4	
US	16.5	37.9	78.8	36.2	3.3	4
Canada	11.4	20.1	66.7	24.4	0.1	7
Japan	21.4	5.8	55.7	12.8	4.2	11

Notes:
(1) subscribers per 100 inhabitants;
(2) per 1000 inhabitants;
(3) mean rank of previous 5 columns;
(4) weighted mean of countries above.

Source: ITU World Telecommunications Development Report, 1998.

4 EUROPE'S PERFORMANCE IN ICT

4.1 Technological Specialization

This section will look more specifically at European performance in ICT production, starting with technological capabilities. We use the same data on US patents as used in the previous section to measure the degree of technological specialization in ICT between countries, making use of the fact that many non-

US-based firms take out patents in the USA. The latter property enables us to assign patents taken out in the US to the countries concerned. Our main indicator is the so-called Revealed Technological Advantage, which is defined as the share of a patent class in a country's overall patenting divided by the share of this patent class in total US patenting. A value of one indicates a neutral position, values above (below) one indicate positive (negative) specialization.

Table 5.5 summarizes the RTAs of European and other major OECD countries in three periods: 1969–74,1979–84 and 1989–94. In terms of the level of specialization, Japan shows a clear and consistent trend of values above one. Europe is consistently below one, and the USA fluctuates around a neutral value (but with a somewhat decreasing trend over the most recent years).[2] Over time, these patterns seem to be reinforced, that is, Japanese specialization in ICT is growing, and Europe's RTA value declining. In Europe, the only countries that have a positive specialization in the last period are the Netherlands (mainly due to Philips, see also Chapter 4), and Ireland, the main success story of the EU (due to the inward flow of FDI in ICT-related industries, see also Chapter 4). France and the UK have RTA close to unity, Germany is clearly below one, and declining over time.

Table 5.5 RTAs in ICT by country, based on US patents, 1969–74, 1979–84, 1989–94

Country	1969–74	1979–84	1989–94
Europe	0.86	0.84	0.73
US	1.02	0.98	0.90
Japan	1.40	1.47	1.53
Austria	0.73	0.48	0.44
Belgium	0.57	0.44	0.47
Germany	0.75	0.73	0.59
Denmark	0.54	0.56	0.33
Spain	0.48	0.28	0.42
Finland	0.56	0.48	0.75
France	1.02	1.12	0.93
UK	0.94	0.92	0.91
Greece	1.07	0.49	0.64
Ireland	0.86	0.49	1.12
Italy	0.68	0.70	0.54
Netherlands	1.45	1.49	1.29
Portugal	1.10	0.00	0.24
Sweden	0.73	0.62	0.63

Source: Calculations on the SPRU/OTAF database on US patents.

What makes the European picture rather bleak is not so much the low level of specialization (because RTA is an indicator of relative rather than absolute advantage), but the negative trend. The RTAs of the European countries are all decreasing apart from those of Finland (mainly due to Nokia) and Ireland, which shows a considerable improvement in the 1990s. Thus, in times when developments in ICT are at the forefront of world technological development, Europe's technological specialization pattern is moving away from these activities, rather than closing in on them.

A more disaggregated look at the data at the level of individual patent classes leads to a picture that is somewhat more favourable as regards European performance in some cases (we do not document these data in order to save space). There are a few areas, especially telecommunications equipment, in which some European countries are well positioned, but in the majority of the major ICT classes Europe lags behind and the trends are usually negative, with some isolated exceptions such as Ireland. It has to be kept in mind, however, that the patent data do not capture technological efforts in fields related to the non-hardware parts of ICT, such as software and services.

4.2 International Trade and Production

As is obvious from the introduction and section 2, the ICT sector can hardly be identified as a historically constant phenomenon. Its dynamics are so rapid that it is difficult to measure its contribution to international trade or production. However, it is clear that all developments in ICT originate historically from a limited number of activities bundled in hardware-producing industries such as semiconductors, computers and telecommunications. While in today's context, this group of three sectors is in no way adequate to measure ICT, it may serve as a broad guideline for a historical perspective on trade performance of the three major blocs: Europe, the USA and Japan. Table 5.6 therefore presents an overview of market shares of these three blocs on world export markets (including intra-EU trade).

The table sketches a by now familiar picture. Europe dominated ICT-related trade in the early 1960s, although this was largely based on intra-European trade at the time. The 35 years afterwards show Japan as a rising star, and Europe and the USA both as losers of market shares. Losses in Europe were much more dramatic than in the USA, however, with for example a loss of half the market share in semiconductors, and about one-third in the two other classes. This confirms the picture already sketched in the Introduction and Chapter 2. Japan's rise was especially spectacular in semiconductors, that is, the intermediate products from which most other ICT-related equipment is built.

Given the rapid technological and economic developments in the ICT fields, such a highly aggregated and narrow definition no longer seems adequate,

Table 5.6 Long-run market shares in core ICT hardware exports

Product group	1961	1965	1969	1973	1979	1984	1988	1994
EU14 (including intra-EU trade)								
Computers & peripherals	56	58	53	56	49	41	44	44
Semiconductors	60	62	51	49	49	30	35	31
Telecommunications equipment	68	64	55	58	57	41	39	45
Total	63	62	53	55	52	38	41	41
Japan								
Computers & peripherals	2	2	7	16	10	18	23	25
Semiconductors	7	8	7	11	24	34	37	40
Telecommunications equipment	5	8	12	17	21	34	39	27
Total	4	6	9	15	17	26	30	30
US								
Computers & peripherals	34	34	34	21	35	37	29	26
Semiconductors	30	27	39	37	25	33	26	26
Telecommunications equipment	21	22	24	18	18	19	16	23
Total	27	26	31	24	27	31	25	25

Source: The IKE Trade Database.

however. Both in the field of hardware manufacturing and services, more detailed data are necessary. Table 5.7 provides such a more detailed view for 1994 (most recent data available at the time of writing). Here, the ICT hardware field has been divided into ten subcategories. Together, these ten classes comprise a set of goods that is larger than the one for 1994 in Table 5.6, which reflects the broader perspective taken for the most recent period.

Table 5.7 gives export market shares for the major world economies and 14 EU countries and Norway.[3] The first row in the table gives the percentage distribution of total trade value over the ten classes. It is clear that computers (plus peripherals) comprise the largest category, with a share of almost one-half. Integrated circuits (ICs) and microassemblies (that is, the building blocks of much ICT equipment) make up the second largest class, with telephone equipment (excluding mobile) in third place. The other two categories with substantial (larger than 5 per cent) market shares are mobile phones, and printed circuits and simple semiconductors.[4]

The computers sector is clearly dominated by the USA and Japan. A further subdivision into computer systems and peripherals (not documented) shows that Japan is stronger in the latter, and the USA in the former. Intra-EU trade in computers comprises a substantial part of total world trade (one-quarter),

Table 5. 7 European performance in international trade in ICT-related hardware classes, 1994

Market shares (%)	Computer and peripherals	Photocopying apparatus	Telephone (excl. mobile) and fax app.	Radio transmission and mobile phones	TV cameras	Radar, reception app. for radio telegraphy	Electromedical apparatus	Printed circuits and simple semicond.	ICs and microassemblies	Piezo-electric crystals	Total
Market size (% of total)	47.7	2.5	10.4	5.3	2.0	1.3	1.9	5.1	21.9	1.9	100
Canada	3.2	0.0	6.6	3.7	0.2	2.5	0.6	4.0	2.6	0.9	3.3
US	25.9	4.8	17.9	30.4	1.7	35.3	38.7	23.0	30.1	18.6	25.3
Japan	25.1	50.2	21.6	18.3	86.3	25.6	22.6	33.2	35.3	67.9	29.7
EU14 (ex. intra-EU)	17.6	11.4	37.9	32.4	3.5	22.1	23.3	21.6	17.2	7.3	20.1
Intra-EU14 trade	26.4	33.3	12.8	12.4	7.9	10.6	11.9	16.4	14.4	4.0	20.0
Austria	0.4	0.2	1.3	0.1	0.2	0.2	0.8	1.8	0.8	0.6	0.6
Belgium/Luxembourg	1.2	1.0	3.9	0.5	0.7	0.7	0.3	1.1	0.6	0.3	1.3

Denmark	0.8	0.4	0.7	1.5	0.1	0.9	2.7	1.0	0.0	0.2	0.6
Finland	0.7	0.1	2.5	6.8	0.0	0.1	2.2	0.2	0.1	0.0	1.1
France	5.2	6.8	6.8	3.6	1.5	8.8	3.1	6.4	4.7	0.8	5.1
Germany	7.4	8.4	14.5	11.0	3.2	6.1	14.3	10.8	8.1	4.0	8.6
Greece	0.0	0.0	0.2	0.0	0.0	0.0	0.0	0.0	0.0	0.0	0.0
Ireland	5.1	0.0	1.1	0.9	0.4	2.5	0.3	0.7	1.3	0.5	3.0
Italy	3.6	2.4	2.7	1.4	0.3	2.0	1.9	2.4	3.0	0.9	3.0
Netherlands	6.7	20.3	2.5	0.4	3.2	1.6	5.2	4.3	3.9	1.9	5.3
Norway	0.3	0.0	0.6	0.7	0.0	1.2	0.9	0.1	0.0	0.1	0.3
Portugal	0.0	0.0	0.2	0.0	0.1	0.0	0.1	0.0	0.3	0.0	0.1
Spain	1.0	0.1	2.6	0.3	0.2	0.8	0.3	0.8	0.3	0.1	0.9
Sweden	0.6	0.4	6.1	10.3	0.1	1.2	1.1	1.1	0.4	0.1	1.7
Switzerland	0.6	0.1	1.0	1.3	0.2	1.8	1.2	1.4	0.4	0.6	0.7
UK	11.4	4.5	5.5	8.0	1.5	7.7	3.0	7.2	8.0	1.7	8.9

Source: The IKE Trade Database. Shares for individual European countries include intra-EU trade. Sectoral aggregation (SITC rev 3 codes): column (1): 75113, 7512, 752, 75991, 75995, 75997; column (2): 7513; column (3): 76411, 76413, 76415, 76417, 76419; column (4): 7643; column (5): 76482; column (6): 76481, 76483; column (7): 7741; column (8): 7722, 7763; column (9): 7764; column (10): 7768.

but extra-EU exports of the EU14 clearly lag behind the market shares of either Japan or the USA. Within Europe, the UK and Germany are the largest exporters. Also, Ireland and the Netherlands, as small countries, have a substantial market share.

Japan has overtaken both the USA and Europe in ICs and microassemblies, with a market share of 35 per cent. The USA comes second with 30 per cent. Total European trade (that is, extra- plus intra-EU) is slightly above 30 per cent, but almost half of this is intra-EU trade. The share of extra-EU exports of EU countries is around 17 per cent, well below the market share of Japan or the USA. Within Europe, Germany and the UK are leaders, with the Netherlands also performing relatively well.

European performance is much better in the communications related areas. In telephones (excluding mobile), the market share of European countries, including intra-EU trade is almost 50 per cent. In mobile phones, the situation is similar. Moreover, in both sectors, extra-EU exports from EU14 countries are substantial, that is, much larger than either the US or Japanese market share. In mobile phones, the USA (mainly Motorola) is very visible with 30 per cent, Japan has 'only' 18 per cent, but Germany (Siemens) has 11 per cent, Sweden (Ericsson) 10 per cent, Finland (Nokia) 7 per cent, and the UK 8 per cent. The establishment of GSM as a 'European' standard obviously plays a role in European dominance in this area, as will be explored in more detail below. This has kept the Japanese somehow in the periphery, but not the Americans because Motorola bought the Danish company Storno in 1986, and from that platform participated in the GSM process. Motorola is very dominant in OECD exports to, for instance, Latin America (not documented). More recently, the Japanese seem to have entered the field, but this does not yet show up in the 1994 data presented here. Non-mobile telephone equipment is also dominated by Europe; especially Siemens in Germany has a strong position, as has Ericsson from Sweden. Of course, a good deal of political influence is used in decisions on large investment in this area. But still Siemens, Ericsson and Alcatel are competitive to a certain extent *vis-à-vis* the US or Japanese giants. The Netherlands, which was very visible in the computers and IC sectors, is not particularly strong in the telecommunications equipment markets.

Together, these trends clearly underline that Europe is specialized, as well as fairly competitive (in absolute terms) in the telecommunications sector. In light of the convergence trend discussed in section 2, this means that European countries are entering the converging ICT markets from a very specific perspective, namely that of the telecommunications technologies.

The data presented so far almost exclusively refer to manufacturing products, that is, hardware. As was argued in sections 2 and 3, however, services, or more generally, non-tangible products, are becoming an increasingly important part of the ICT market. The measurement problems in this area are, however, formidable (see, for instance, OECD, 1998, for an overview of some of the issues). This is the reason why we are unable to present a coherent and comprehensive picture of European competitiveness in this part of the ICT market. The limited evidence that is available relates to software sales. Comprehensive data on international trade in this field are not available, but OECD (1998) presents a table that lists the 20 leading firms in the world, with their (estimated) sales in 1996. In this top-20, 15 firms are based in the USA. In 1996, IBM was still the world leader, with Microsoft ranking second. There are two Japanese firms in the list: Hitachi at rank 3, and Fujitsu at rank 6, and there are three European firms: SAP from Germany at rank 7, Bull from France at rank 8, and Siemens-Nixdorf from Germany at rank 11.

These data show the dominance of the USA in the world software market, and seem to point out that European (and Japanese) firms are lagging behind. However, this does not take into account the role of smaller firms, often based in the smaller EU countries (such as Baan from the Netherlands). Thus, the overall picture for Europe may be more positive than this top-20 of world software producers seems to show. It is obvious, however, that data for the smaller firms would not turn around the picture completely, so that it seems justifiable to conclude that Europe's position in software does not give rise to more optimistic conclusions than is the case in hardware.

Overall, the data in this section point to a European position in ICT hardware that has been weakening ever since the early 1960s. The USA has been a leader in the field over the whole period, with Japan catching up from a backward position, overtaking Europe, and now challenging the USA in some fields. In some areas, the European position is still favourable, such as telecommunications equipment. Thus, from the point of view of technological convergence in the ICT field (see section 2), Europe's position is one that is clearly rooted in one of the converging streams: telecommunications. The European position in software is similarly weak.

4.3 Europe and Mobile Communications

As was shown above, European performance in the field of mobile communications has been relatively favourable. This section will argue that one important reason for this favourable position lies in the decisive role of the development of technical standards, which was a joint effort between (public) telecomms service providers, telecomms regulators and private firms. In other words, the case of European mobile communications shows how public–private cooperation at the institutional level may be a decisive factor in enhancing the competitiveness of European firms.

Mobile communications has been an area of unprecedented growth in the 1980s and 1990s. The growth prospects, measured in terms of increase in the number of users as well as the intensity of traffic on the mobile networks (existing subscribers using new facilities), are very positive in almost any country in the world. Although demand for mobile services has developed to a universal phenomenon, development and, less so, production of the equipment are concentrated in a rather small number of multinational firms, which coexist with dozens of small product development-oriented firms.

The antecedents to present day mobile phone systems were closed land mobile networks used by, for instance, police and transport companies. In the 1960s the three largest equipment producers in this field were General Electric and Motorola of the USA and the Danish firm Storno. The core ingredient of the public mobile phone systems introduced in the 1980s is the cellular concept invented by Bell Laboratories (now part of Lucent Technologies) and introduced in 1979. In 1981, the more ambitious Nordic Mobile telephone system (NMT) was introduced, which operated across Sweden, Denmark, Norway and Finland. NMT was a concerted effort between the national telecommunications providers (then public monopolies), public regulators and a few private firms – first Storno and the Swedish Ericsson and later also Nokia of Finland.

This targeted effort led to a sharp increase of the subscriber penetration ratios (number of users per capita) in the Nordic countries compared to the USA and the rest of the world from the early 1980s onwards, as documented before in Table 5.4 (for historical accounts see Johnston, 1998; Garrard, 1998; Dalum, Fagerberg and Joergensen, 1988, Ruotto, 1998). The high levels of user intensity were an important demand side incentive behind the phenomenal growth of Ericsson and Nokia in this field in the 1980s. The third worldwide player, Motorola, dominated the US market (based on a US standard) but was also present on the Nordic market from 1986, through acquisition of Storno in Denmark. Although the final outcome of this process was a common standard, the NMT, it is important to emphasize that the initiative was taken by engineers at the four Nordic PTTs (Post, Telegraph and Telecommunication Authorities), which were combined regulators and public monopoly service providers. There has for many decades (if not the entire twentieth century) been a tradition of cooperation between the Nordic countries in a whole range of different fields, for instance, in all sorts of academic and more practical fields (history, economics, technical disciplines, etc.) at annual meetings, conferences, in smaller work groups, etc. The NMT system emerged out of such an R&D dominated telecommunications environment, which again had close relations to the private sector firms. In this case, the PTT engineers appear to have been the most innovative part initially. The NMT system, the direct ancestor of the GSM system, was thus the result of visions among innovative engineers in public research laboratories.[5]

Usually the cellular systems mentioned so far are grouped as first generation, based on analogue technologies. The successful Nordic case was taken over by the EU in the mid- to late 1980s. Technologically it became evident that future systems could be based on fully digitized principles. European companies and telecomms authorities/monopoly service providers (PTTs) initiated the first discussions of what later became the now dominant GSM system. From around 1987/88 the EU took over the standardization coordination in a new institution, the European Telecommunications Standards Institute (ETSI) – a move which was published at the time as the 'high tech flagship project of Europe 1992'. The ETSI/GSM project has been of obvious importance for the European presence in the equipment industry for mobile communications. Even within the GSM framework, the Nordic countries preserved their lead by, for instance, being the first countries to test and introduce the new system, resulting in their present-day leader position in the worldwide subscriber penetration 'league'.

At the industrial level, the three dominant firms in the 1990s have been Ericsson, Nokia and Motorola followed by Siemens (Germany) and, more recently, some of the Japanese multinationals, such as Sony and Matsushita (with the brand name Panasonic). The important point to notice, however, is that the Japanese and the Americans have lagged behind at the institutional front. Both countries have lacked the capacity to create dominant standards, even domestically, with serious implications for competitive strength for US and Japanese firms.

The process of creating a third generation standard, which will integrate voice and multimedia services, is now already at a fairly advanced stage. The new standard (UMTS, established in January 1998) has also been based on a European initiative, coordinated through ETSI. UMTS is in fact a compromise between two competing consortia of multinational equipment producers and some of the dominant telecomms service providers. The majority group was headed by Ericsson, Nokia, the former Japanese monopoly service provider NTT DoCoMo, the dominant Nordic service providers (the old monopolies) and also backed by the US Lucent Technologies (part of former AT&T, including Bell Laboratories). The minority group contained for instance Motorola, Siemens, Bosch, Nortel (Canada), Sony and Alcatel. Nearly all the major multinational telecomms hardware companies were thus involved. There is, however, one major conflict still not solved. The UMTS system is based on wideband CDMA (Code Division Multiple Access). The US firm Qualcomm has developed CDMA for use in second generation systems. Qualcomm asserts that some of the technologies selected for use in the UMTS standard are protected by Qualcomm patents. The case was due to be taken to court in the USA in 1998 – formally between Ericsson and Qualcomm; but the entire group of large telecomms multinationals apparently backs the Swedish firm via ETSI.

Still, the involvement of European companies in the new standard is large, and hence it seems fair to conclude that at the institutional level, the EU has managed to stay in a pivotal position. What may, on the other hand, cause worries in Europe is that the first user of the UMTS system is not a European service provider. It will be the Japanese NTT DoCoMo, which will launch the first UMTS service around 2002. As is obvious from the Nordic case of NMT, it may be a disadvantage for future development that the pioneer market for the third generation system will be Japan and not Europe, although Ericsson and Nokia, at least initially, have been the dominant equipment developers in the Japanese context. But there is a risk that this pattern may change in favour of Japanese companies such as Matsushita (Panasonic) and Sony.

5 CONCLUSIONS AND DISCUSSION

The empirical evidence presented in this chapter brings out three clear trends. First, ICT-related goods and services are spreading rapidly throughout the world economy, although at different rates in different countries. These new technologies have a pervasive impact on all parts of the production structure, as well as on final consumption. Second, the competitiveness of European producers in ICT seems to be weak in general. The main exception to this latter trend is European performance in telecommunications, especially mobile communications, where the Nordic countries in particular are strong. Third, the diffusion of ICT-related products takes place at very different rates in different European countries; in particular, there seems to be a strong north–south dividing line in Europe. Compared to the USA and Japan, the Nordic countries show high adoption rates, the large western European countries low or average rates, and the southern European countries (very) low rates.

This raises a number of questions. Why is European performance in ICT production weak, and what makes the telecommunications sector the main exception to this? What causes the high adoption rates of ICT products in Nordic countries, and, conversely, what causes the low rates in southern European countries? What is the connection between the two phenomena (adoption and competitiveness)? And, finally, what can be done about this by European policy-makers? These are obviously more complex questions than can be solved in this chapter, but we will conclude by some discussion of possible answers.

One possible answer to the question of what causes the lack of European competitiveness in the field of ICTs, would be a lack of technological competitiveness. The empirical evidence presented here indeed points out that European firms are becoming less and less specialized in ICT. However, given the interpretation offered in the introduction of ICT as a new techno-economic paradigm, such a purely technology-based explanation does not seem to be very convincing. Instead, the view that technological systems such as ICTs are rooted in a host of societal, economic, technological and institutional factors suggests that there are a number of factors that jointly explain the lagging European performance.

One such factor may be the organization of markets. The process of European integration has traditionally been aimed at creating a 'level playing field', in which free competition ensures efficiently operating firms. Saxenian (1994) argues that in the case of the computer firms located in the so-called Route 128 area in the USA, monopolistic structures may indeed have been a major brake on economic and technological performance. However, her discussion of the case of Silicon Valley shows that a more ideal system may require more than just free competition, that is, a (local) network of small firms that cooperate in a wide range of activities, such as R&D and production (see also Castells, 1996). In other words, creating a European-wide 'level playing field' by means of an active competition policy may simply not be enough to create an environment in which competencies in the field of ICTs can prosper.

Another factor on the institutional side may be the availability of human resources. Both the development of ICT and the applications of ICT in all fields of production require sufficient human capital. The pervasive application and diffusion of a major new technology requires millions of newly skilled people and the retraining of millions of others. The speed and effectiveness with which a new institutional and social framework can be developed must necessarily vary with each country's institutions and the characteristics of each new technology. Nowhere is this more apparent than in the case of software skills, where the shortages persist after more than 30 years, and numerous accounts demonstrate how the shortage of skills has retarded growth or even completely frustrated the achievement of productivity gains.

The example of mobile communications in Europe (and in the Nordic countries in particular) suggests that the production and consumption side of ICT may well develop in close synergy. When a large enough initial market is created, learning effects on the production side may well provide local producers with a great advantage, and enable them to develop strong technological and market competencies which are a solid basis for exporting. The low rates of adoption of various ICT products in parts of Europe, especially southern Europe, but also in some respects in the large European countries such as the UK, Germany and France, suggest that such a 'virtuous circle' has been lacking for the single European market in the 1990s. The analysis here did not provide an explanation for the different adoption rates, but it seems obvious that differences with respect to culture, education, life styles, etc., as well as economic development in general are a large part of such an explanation. This implies that, although tariff and non-tariff barriers to trade have largely been removed after 1992, the European market for services and products in the field of ICT is still very heterogeneous due to different levels of user preferences and user sophistication in different parts of Europe.

To remedy such a situation, the example of mobile communications shows that a large role can be played by technological standards. Obviously, this is primarily an issue for the market parties, but European policy-makers can and should have a facilitating role in this process (see also the case of HDTV, now applied in so-called digital video broadcasting). However, technological standards alone will not solve the situation of a European market divided by cultural differences and different life styles. An active policy aimed at homogenizing the market in this respect may well be very effective. Such a policy should of course not be aimed at ironing out such differences between the individual countries, but rather at providing the adequate local incentives and prerequisites for more rapid adoption of ICTs (see Castells, 1996). Such policies have been suggested for European involvement in the Internet, for example by stimulating the supply of public services through the net.

NOTES

* With contributions from Keld Laursen and Umit Efendioglu. The authors thank Rajneesh Narula for useful comments and suggestions.

1. Note that the definition of ICT-related patent classes is open to debate. We opted for the classes in the table, some of which obviously have a broader scope than just ICT.
2. Because of a 'home-market' effect, US firms still dominate total US patenting, which partly explains the little pronounced specialization pattern of the USA. The large number of patents assigned to US firms implies that this country dominates the overall sectoral distribution, and hence that its RTA is close to unity.
3. The table gives absolute market shares. For relative performance (specialization indicators), see Chapter 2.
4. 'Simple semiconductors' includes goods such as resistors, diodes (including leds), elcos and the like.
5. Another example of successful public–private cooperation in Nordic telecomms R&D was the development of one of the first fully digitized public switches for wired telephone networks in 1977; the AXE system by Ericsson. The AXE system has been an important competitive edge for Ericsson in the 1980s and 1990s and also the technical foundation for Ericsson's leading position in public switches for mobile communications. The switch was developed 1970–77 at a joint venture company, Ellemtel, formed by Ericsson and the then Swedish monopoly telecom provider, Televerket (now Telia) with 50 per cent each.

REFERENCES

Castells, M. (1996), *The Rise of the Network Society*, Oxford: Blackwell.

Dalum, B., J. Fagerberg and U. Joergensen (1988), 'Small Countries in the World Market for Electronics: The Case of the Nordic Countries', in B.-A. Lundvall and C.

Freeman, (eds), *Small Countries Facing the Technological Revolution*, London: Pinter.

Duysters, G. (1996), *The Dynamics of Technical Innovation*, Cheltenham: Edward Elgar.

Garrard, G.A. (1998), *Cellular Communications. Worldwide Market Development*, Boston: Artech House Publishers.

Johnston, W. (1998), 'Europe's Future Mobile Telephony System', *IEEE Spectrum*, Vol. 35, No. 10.

Landes, D. (1969), *The Unbound Prometheus. Technological Change 1750 to the Present*, Cambridge: Cambridge University Press.

Meliciani, V. and R. Simonetti (1998), 'Specialization in Areas of Strong Technological Opportunity and Economic Growth', in Eliasson, G. et al. (eds), *Microfoundations of Economic Growth. A Schumpeter Perspective*, Michigan: University of Michigan Press.

Mokyr, J. (1990), *The Lever of Riches: Technological Creativity and Economic Progress*, New York: Oxford University Press.

Mowery, D.C. and N. Rosenberg (1989), *Technology and the Pursuit of Economic Growth*, Cambridge: Cambridge University Press.

OECD (1998), 'The Software Sector: A Statistical Profile for Selected OECD Countries', unclassified document of the Committee for Information, Computer and Communications Policy, Directorate for Science, Technology and Industry.

Perez, C. (1983), 'Structural Change and the Assimilation of New Technologies in the Economic and Social Systems', *Futures*, **15**: 357–75.

Ruottu, A. (1998), *Governance Within the European Television and Mobile Communications Industries, 1970–96. PAL Plus and GSM – A Case Study of Nokia*, DPhil thesis, University of Sussex.

Saxenian, A. (1994), *Regional Advantage. Culture and Competition in Silicon Valley and Route 128*, Cambridge, MA and London: Harvard University Press.

Schumpeter, J.A. (1939), *Business Cycles. A Theoretical, Historical and Statistical Analysis of the Capitalist Process*, New York: McGraw-Hill.

Soete, L. and N. von Tunzelmann (1987), 'Diffusion and Market Structure with Converging Technologies', Research memorandum, University of Linburg, Maastricht.

6. Lack of regional convergence

Aadne Cappelen, Jan Fagerberg and Bart Verspagen

1 INTRODUCTION

Greater equality across Europe in productivity and income has been one of the central goals for the European Community since the early days of European economic integration. And for a long time this was achieved. As shown by several studies,[1] differences in GDP per capita between European regions declined steadily from the early 1950s onwards. However, more recently this process seems to have slowed down considerably, and perhaps come to a complete halt (Fagerberg and Verspagen, 1996).

What are the reasons behind these developments? Economic analyses of differences in growth across countries or regions have been based on two competing perspectives. The first, based on the traditional neoclassical theory of economic growth (Solow, 1956), is based on the assumption that technology is a public good, available to anyone free of charge. This perspective puts the emphasis on capital accumulation as the main vehicle for reducing differences in productivity across countries or regions. Moreover, this is assumed to happen more or less automatically, as long as markets are allowed to work freely. The other, competing, perspective puts the main emphasis on innovation and diffusion of technology as the driving force behind differences in growth. This perspective is based on a totally different view on technology, emphasizing its public as well as private character, and the complementarity between technology and other factors that take part in the growth process. Technological know-how is described as a rather local affair, embedded in firms, organizations, networks and so on, dependent on specific assets of various types that are not easily forthcoming. This leads to the hypothesis that without the ability to develop such assets, countries or regions are likely to fall behind rather than catch up.

Previous research has shown that the predictions of the public good model do not fit regional growth very well (see, for instance, Sala-i-Martin, 1996) and that a more comprehensive framework, taking into account a wider range of factors, is necessary in order to accommodate the facts. This is what we wish to do in this chapter for the European case. We start in section 2 by laying out some stylized facts on regional growth in Europe from 1960 onwards. On this basis we proceed in section 3 to an econometric analysis on how these facts can be explained, allowing for a broad range of causal factors as well as the possibility that the dynamics differ across various types of regional 'clubs'. Section 4 broadens the analysis by taking into account the possible impact of changes in EU regional support in the last decade. Finally, in section 5 we return to the questions of what all this has to say for regional convergence in Europe in the future, what the lessons for policy are and what the future research agenda in this area should look like.

2 REGIONAL CONVERGENCE IN EUROPE: THE STYLIZED FACTS

In this section we take a fresh look at regional disparities in GDP per capita in EU countries. We extend the data to the most recent ones available (1995) at the time of writing. For all countries except the most recent members of the EU (Austria, Finland and Sweden), EUROSTAT provides data on regional income data in purchasing power standards (PPS) values from 1980 to 1995. We link these data with earlier data as consistently as possible in order to study fairly long historical series on dispersion within and between EU countries. In addition we provide some evidence on regional dispersion of labour productivity.

Our data on regional per capita GDP levels cover the period 1960 to 1995. The data for 1960 are based on Molle (1980). The data for 1970 to 1980 are based on various issues of the yearbook of regional statistics from EUROSTAT. More recent data are from the REGIO database.[2] All data are measured in PPS (purchasing power standard). As a measure of dispersion, we use the standard deviation of the logarithm of regional GDP per capita.[3]

Table 6.1 presents the trends for two country groupings, EU9 and EU12. The difference between these two is that the latter also includes the three southern countries that joined the Community in the 1980s; Greece, Portugal and Spain. For each grouping we present two measures of dispersion, *regional* and *national* standard deviation. The former measures the dispersion across regions in Europe, irrespective of which country the region belongs to. The latter measures the dispersion between country aggregates (that is, means over regions).

The table shows that a significant decrease in both regional and national dispersion took place in EU9 between 1960 and 1980. Hence, Europe appeared to be on a path of convergence. After 1980, this trend continued at the national level (that is, between countries), although at a slower pace. The latter is not so surprising, given that the level of disparity was rather low already in 1980. What is more surprising, however, is that at the regional level, for which the level of disparity in 1980 was much higher, convergence ceased after 1980. These findings are in accordance with the figures presented by Molle and colleagues, using a different methodology (Molle, 1980; Molle and Cappellin, 1988).[4]

Table 6.1 Dispersion of GDP per capita in EU regions and countries 1960–95

	1960	1970	1980	1995
Regional standard deviation EU9	0.343	0.274	0.202	0.205
National standard deviation EU9	0.263	0.233	0.184	0.168
Regional standard deviation, GDP per worker, EU9	0.306	0.211	0.145	0.153
National standard deviation EU12	0.459	0.342	0.328	0.258
Regional standard deviation EU12			0.279	0.255
Share of total regional dispersion in EU9 due to:				
Between country dispersion in per cent	64	79	52	41
Within country dispersion in per cent	36	21	48	59

Notes:
All figures based on PPS. The regional disaggregation is more detailed from 1980 and onwards and this biases the figures before 1980 for EU9 downwards. The productivity figures for EU9 do not include Denmark and Ireland and are based on the NUTS-I level for most countries, as employment data at the NUTS-II level were not available. The share of total dispersion is based on a standard decomposition of the total variance of log GDP per capita in between country and within country effects.

As can be seen from Table 6.1, the level of dispersion in GDP per employed person is lower than in GDP per capita. This implies that high-income regions are rich not only because they have high productivity but also because a larger share of their population is working. However, the change in the dispersion of GDP per employed over time is quite similar to that of GDP per capita. This implies that in the present context, the use of GDP per capita as a proxy for productivity is relatively unproblematic.

The larger group (EU12), including the southern entrants, deviates from the EU9 experience by showing both regional convergence and national convergence between 1980 and 1995. Furthermore, the reduction in national disparity (that is, between country aggregates) is much stronger than for EU9.

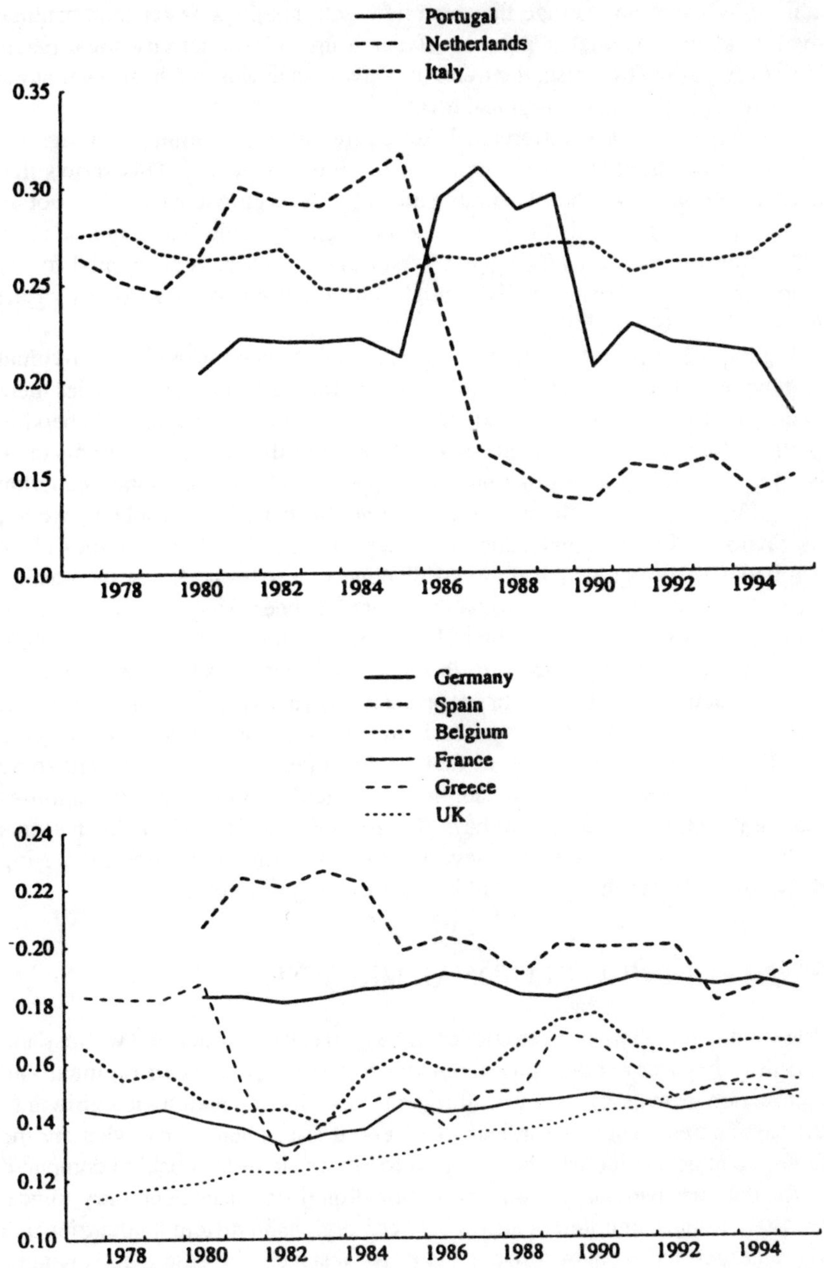

Figure 6.1 Dispersion in regional GDP within European countries

This shows that on average the three new entrants grew faster than original members, that is, that a process of catch-up in productivity took place. However, also in this case, the trend of convergence was much stronger at the national level than at the regional level.

We thus find a clear difference between trends at the national and regional levels, with much less convergence at the regional level.[5] This shows that although countries in the EU at large are still converging to each other, not all regions benefit from this process to the same extent. This is also indicated by the increasing (since 1970) share of total disparity that is accounted for by within-country disparity in EU9, rather than between country disparity (last lines in Table 6.1).

Figure 6.1, which reports data for regional dispersion within individual member countries, further illustrates this pattern.[6,7] For all EU9 countries there is either little change, or a tendency towards increased regional dispersion (notably for the UK and to some extent Italy[8]). For the three new entrants there is a lot of variation over time, but there appears to have been some reduction in regional disparity in all three countries over the period as a whole. However, for Spain and Portugal this reduction is very small, and for Greece it took place prior to 1982, after which regional dispersion has been slowly increasing. Hence the conclusion seems to be that there has been very little convergence between regions within individual EU member countries since the early 1980s.

The evidence brought forward in this section confirms that the process of convergence in GDP per capita (or productivity) across European regions has indeed slowed down in recent years. However, this is not caused by diverging trends between EU countries. In fact, at the country level there is still some evidence of convergence. This applies in particular to the southern countries that entered the community in the early 1980s. Rather, this slow-down seems to be related to the lack of convergence – and in some cases increasing dispersion – between regions within most EU member countries.

3 EXPLAINING REGIONAL GROWTH

The finding that there is more convergence between countries than within them certainly begs new questions about the relationship between regional and nationwide growth. Clearly not all regions benefit from nationwide growth to the same extent. There are at least two central issues here. First, what are the factors that determine whether a region forges ahead or lags behind compared to the country average? Second, conditional on differences in country trends, are there important commonalities between regions in different countries that share certain basic characteristics? Is it, for instance, the case that, say, agricultural regions have much in common, independent of which country they belong to? These are issues that we would like to explore in the following.

Any explanation of growth differences needs theoretical underpinning. Our point of departure is the 'technology gap theory' of growth (Fagerberg, 1987).[9] As in other theoretical frameworks, it is assumed that innovation, diffusion and technology fuel growth. However, although technological activity gives rise to positive externalities, technology is not assumed to be a public good in the sense that it is equally available to everybody free of charge. On the contrary, it is argued that successful adoption of new technology is generally costly. Typically, it requires a host of complementary factors of the sort that Abramovitz (1994) classifies under the terms 'social capability' and 'technological congruence'. Hence, following this perspective, regional growth may be seen as the outcome of three sets of factors:

- innovation activities in the region;
- the potential for exploiting technologies developed elsewhere (diffusion); and
- complementary factors affecting to what extent this potential is realized.

There are two major problems in applying this perspective. The first has to do with finding indicators of innovation and the potential for diffusion, the second with identifying and measuring the 'complementary factors'.[10] For innovation we use R&D intensity, defined as business enterprise R&D personnel as a percentage of total employment. We expect a positive impact of this variable. For diffusion potential we use, as customary in the literature, the initial level of GDP per capita in the region (log-form, 1980). The higher this level, the smaller the scope for imitating more advanced technologies developed elsewhere. Hence, the expected impact of this variable is negative. Regarding complementary factors, there are many candidates that can be defended theoretically, from variables related to various types of investment in people (that is, education), infrastructure and physical capital to structural factors of various sorts. However, data are extremely scarce, especially among the former.

The 'complementary' variables that we were able to take into account include:

- physical infrastructure (kilometres of motorway per square kilometre);
- population density (the number of inhabitants per square kilometre);
- industrial structure (the shares of employment in agriculture and industry, respectively, in total employment);[11]
- long-term unemployment (that is, duration of more than one year, as a share of the total labour force); and
- growth rate of the population in the region.

Among these, we would expect the first two to have a positive impact on technology diffusion, since both a more developed infrastructure and a higher population density increase the profitability/reduce the cost of introducing new technology. Regarding industrial structure, it is one of the standard results in the existing empirical literature on regions that industrial structure matters. In particular, a high reliance on agriculture has been shown to be detrimental to regional growth (Fagerberg and Verspagen, 1996), among other things because of low technological opportunities, and slow growth of the market. On the share of 'industry' in total employment the expectations are less clear. Traditionally this sector – particularly manufacturing – has been regarded as an 'engine of growth' (Kaldor, 1967). However, technological progress in recent decades has been more geared towards services than industry and many traditional industries have been characterized by slow growth. Finally we include the level of unemployment as a possible complementary factor. We interpret this as a measure of the cohesion of the broader social and economic system in the region. The higher the share of the labour force that is excluded from work on a long-term basis, the less well this system works. Hence it is an indicator of institutional failure, and as such it might be expected to have a negative impact on growth.[12] For instance, it may hamper inflows of risk capital and qualified people, and encourage outflows, as empirical research in this area indeed suggests (Fagerberg, Verspagen and Caniëls, 1997). Long-term unemployment also leads to deprecation of skills and lack of learning by doing in parts of the workforce. In preliminary estimations we also included population growth, but it never had any significance, and was consequently dropped. All explanatory variables included in the regressions are for 1985, except unemployment, which due to data availability refers to 1987. The dependent variable, growth of GDP per capita, is the mean growth rate over the years 1980–94.

Table 6.2 presents the results. The basic model (first column) explains about 40 per cent of the total variance. When country dummies are added (third column), as is customary when pooling regional data for different countries, the explanatory power increases quite a bit. Furthermore, as shown in the bottom row of the third column, an F-test suggests that these dummies are jointly significant. However, by inspection it turns out that most of these dummies are of little numerical importance. The exceptions are Spain and Portugal, of which the former grows significantly faster, and the latter significantly slower, than the average.

The results support the basic hypothesis of our approach. Both innovation (R&D) and a large potential for diffusion contribute positively to regional growth when the impact of other, conditioning factors is taken into account. Among the latter, all but physical infrastructure receive some support. The evidence suggests that a high reliance on agriculture hampers growth, as does a high share of the workforce in industry. Hence the regions that have been able to exploit the potential for diffusion to the fullest extent are those that are specialized in services. This is consistent with the view that the effects of the

ICT revolution have been more widespread in services than in other areas. For the two remaining conditioning variables, population density and unemployment, the results do to some extent depend on whether country dummies are included or not. Although the signs are as expected in both cases, the estimated coefficients are not significantly different from zero in the latter case. This is especially evident in the case of unemployment. The latter has to do with the fact that long-term unemployment is especially high in Spain. Since Spain is a fast-growing country despite these problems, the variable loses its explanatory power when country-specific effects are not included.

Table 6.2 Basic results. One or several growth clubs?

Variables	Model (A)	*t*-stat.	Model (B)	*t*-stat.	Test for split (A) F(7,91)	Test for split (B) F(7,82)
Longitude					2.02	1.86
Constant	0.1954	(6.17)	0.2386	(4.88)		
GDP per capita	–0.0432	(5.43)	–0.0535	(4.25)	2.56 **	1.95
Agriculture	–0.0375	(4.25)	–0.0398	(3.99)	3.49 *	2.53 **
Industry	–0.0358	(3.71)	–0.0386	(4.10)	2.77 **	2.45 **
Infrastructure	0.0001	(0.55)	0.0003	(0.84)	1.87	1.86
Unemployment	–0.0002	(1.05)	–0.0010	(3.54)	4.52 *	2.15 **
Population density	0.0006	(0.91)	0.0017	(1.96)	2.35 **	1.62
R&D	0.0027	(1.54)	0.0038	(2.73)	3.47 *	3.97 *
Belgium			–0.0009	(0.36)		
Germany			–0.0003	(0.13)		
Greece			–0.0025	(0.67)		
Spain			0.0084	(4.01)		
France			–0.0012	(0.74)		
Italy			0.0025	(1.42)		
Netherlands			–0.0027	(0.82)		
Portugal			–0.0072	(1.98)		
UK			0.0003	(0.19)		
R^2	0.40		0.55			
(R^2)	(0.36)		(0.47)			
N	106		106			
Test for inclusion of dummies F(9,98)			3.29 *			

Notes:
Absolute, heteroscedasticity consistent *t*-statistics in brackets.
* and ** denote significance of *F*-test at the 1 and 5 per cent level, respectively.
R^2 in brackets adjusted for degrees of freedom.

The second question we wish to address is the extent to which regions differ according to other criteria that may cut across country borders. Is it, for instance, the case that the dynamics of, say, regions with high unemployment differ from that of other regions in important respects? We test this using a method developed by Durlauf and Johnson (1992), and subsequently applied by two of us on European growth (Fagerberg and Verspagen, 1996). The method consists of sorting the observations in increasing (or decreasing) order of a control variable, splitting the sample at each observation *j* and running two separate regressions for the first *j* observations and the last *n–j* observations (where *n* is the total number of observations). The final step is to pick the combination of control variable and *j* that yields the highest explanatory power.[13] As control variables we include all the explanatory variables in our model and, in addition, a variable reflecting the location of the region on a north–south axis (*longitude*).[14] The latter is introduced to test for the commonly held view (see, for instance, Neven and Goyette, 1995) that there is a clear division between the dynamics in the north and south of Europe.

The two last columns in Table 6.2 report F-tests for the significance of these splits, the first without – and the second with – country dummies included in the regression. The results are broadly the same, although – as could be expected – there is more evidence in favour of splits when country-specific effects are not allowed for. In both cases the results suggest that the splits based on geography (longitude) and infrastructure do not lead to a significant rise in the explanatory power of the model. Hence there is very little support in the data for a simple north–south dichotomy; rather the results point to a more complex pattern. In both cases the most significant splits are those based on structural factors (agriculture and industry), unemployment and R&D. Among these the split based on unemployment is much less significant when country-specific factors are allowed, for reasons just discussed (the Spain-effect). In fact, the only split that is supported at the 1 per cent level of significance in both cases is the one based on R&D.

Table 6.3 gives some of the characteristics of the most significant regional groupings (variable means). It turns out that there are important similarities between the splits based on R&D, agriculture and industry. In all three cases, we get a split between a small, high-productivity group (20–30 regions) and a larger group with more average/low levels of productivity. The smaller group also has a different industrial structure, with a much more prominent role for services, and a very small agricultural sector. Moreover, these regions also have better infrastructure, higher population density, less unemployment and – above all – much higher investments in R&D. They also tend to be more 'northern', although this is not a defining characteristic in itself, as noted above.

Table 6.3 Characteristics of 'regional clubs'

	R&D		Agriculture		Industry		Unemployment	
Split	High	Low	High	Low	High	Low	High	Low
Number of regions	29	76	85	20	27	78	31	74
Productivity growth	0.014	0 016	0.015	0 017	0.013	0.017	0.018	0.015
Population growth	0.004	0.003	0.003	0.003	0.003	0.003	0.003	0.003
Average longitude	49	43	43	50	47	44	42	46
GDP per capita	8322	5791	6152	7923	7526	6131	5502	6903
Agriculture	0.05	0.18	0.17	0.02	0.07	0.17	0.15	0.14
Industry	0.33	0.29	0.29	0.32	0.40	0.26	0.27	0.31
Services	0.61	0.53	0.53	0.66	0.53	0.56	0.57	0.55
Investment	0.17	0.23	0.23	0.13	0.20	0.21	0.22	0.20
Infrastructure	3.67	1.15	1.32	4.08	2.62	1.58	1.43	2.02
Unemployment	4.83	6.74	6.05	6.89	5.33	6.52	11.11	4.16
Population density	0.62	0.18	0.13	1.05	0.21	0.34	0.47	0.23
R&D	0.93	0.14	0.27	0.77	0.60	0.28	0.15	0.45
Country composition:								
Belgium	0.07	0.01	0.00	0.15	0.00	0.04	0.06	0.01
Germany	0.28	0.03	0.05	0.30	0.26	0.04	0.00	0.14
Greece	0.00	0.17	0.14	0.05	0.00	0.17	0.00	0.18
Spain	0.03	0.22	0.19	0.10	0.15	0.18	0.58	0.00
France	0.28	0.17	0.24	0.05	0.30	0.17	0.03	0.27
Italy	0.07	0 24	0.24	0.00	0.15	0.21	0.26	0.16
Netherlands	0.14	0.00	0.05	0.00	0.04	0.04	0.00	0.05
Portugal	0.00	0.07	0.06	0.00	0.04	0.05	0.00	0.07
UK	0.14	0.09	0.05	0.35	0.07	0.12	0.06	0.12

Table 6.4 reports the working of the model for these regional groupings. In most cases the difference in the estimated coefficient for the same variable across the two subsamples of the split is small and not significant. The main exception is R&D, for which it is positive and significant for the smaller groups, but negative (and significant in two out of three cases) in the larger ones. Thus, while R&D may be an important factor behind differences in growth performance among advanced regions, and between advanced and less advanced regions, it does not seem to be a very efficient tool for regions below a certain threshold of development. This certainly begs new questions about the creation of technological capabilities in such regions.[15] The other exception is industry. Although the impact of specialization in industry is negative for all regional groupings, the estimated impact is much stronger in the larger groups. Hence the regions with low productivity, little R&D, a high share of total

Table 6.4 Explaining regional growth, 'regional clubs'

	High R&D	Low R&D	High agriculture	Low agriculture	High industry	Low industry	High unemployment	Low unemployment
N	29	76	85	20	27	78	31	74
GDP per capita level	–0.0446	–0.0403	–0.0409	–0.0522	–0.0489	–0.0384	–0.0521	–0.0379
	(4.01)	(2.83)	(1.81)	(4.28)	(3.33)	(2.58)	(4.15)	(2.43)
Agriculture	–0.0596	–0.0439	–0.0333	–0.0479	–0.0500	–0.0542	–0.0331	–0.0444
	(6.64)	(3.64)	(2.01)	(4.35)	(4.06)	(4.82)	(3.76)	(3.42)
Industry	–0.0272	–0.0676	–0.0628	–0.0315	–0.0378	–0.1200	–0.0290	–0.0404
	(3.63) *	(4.70)	(2.61)	(3.51)	(4.28) *	(4.29)	(2.38)	(3.02)
Unemployment	–0.0010	0.0009	–0.0009	–0.0009	–0.0011	–0.0010	–0.0007	–0.0059
	(4.65)	(2.34)	(1.42)	(3.52)	(3.78)	(2.93)	(2.94) *	(3.38)
Population density	0.0013	0.0017	0.0926	0.0016	–0.0010	–0.0023	0.0018	0.0120
	(2.50)	(1.10)	(2.11)	(2.03)	(0.27)	(1.51)	(1.93)	(1.50)
Infrastructure	0.0001	0.0003	–0.0065	0.0001	0.0005	0.0015	0.0005	0.0000
	(0.19)	(0.85)	(1.94)	(0.39)	(0.96)	(2.34)	(1.47)	(0.02)
R&D	0.0026	–0.1287	–0.0908	0.0042	0.0041	–0.0095	0.0028	0.0003
	(1.98) *	(4.69)	(3.34) *	(3.25)	(2.50) *	(1.45)	(1.76)	(0.09)
Country dummies	Yes		Yes		Yes		Yes	
$R^2(R^2)$	0.66	(0.56)	0.63	(0.52)	0.63	(0.51)	0.62	(0.50)

Notes:
Absolute, heteroscedasticity consistent *t*-statistics in brackets.
* denotes significance of Wald-test for difference in coefficients across the two subsamples at the 1 per cent level.
R^2 in brackets adjusted for degrees of freedom.

employment in agriculture and so on (in short, the less advanced regions) experience the largest negative growth impulse from industry.

The split based on unemployment is to some extent different from the three other splits (Table 6.3). Again, there is a small group and a larger one, but in this case the regions in the small group are high unemployment regions, of which more than one-half are Spanish. Among the remaining the majority are Italian. On average these regions have low productivity, little R&D, bad infrastructure and a large agricultural sector, as is typical for many 'peripheral' regions. In this case, the only significant difference between the two groups relates to the impact of unemployment on growth, which happens to be much more manifest in the low unemployment regions. However, since this result seems to depend very much on observations from one country (Spain), it is difficult to draw strong conclusions on this basis. We return to this issue in the next section.

The findings in this section support the view that innovation and diffusion of technology are important factors behind regional growth. But innovative efforts in the form of investments in R&D appear to be most efficient in advanced regions. In fact, in technologically less advanced regions, for example regions that do little R&D themselves, there is no evidence suggesting that regions that do more R&D than others fare any better in terms of growth. However, the results confirm that less advanced regions are faced with a considerable potential for growth through exploitation of more advanced technologies developed elsewhere. But the evidence suggests that this potential is seldom fully exploited because of counteracting factors such as, for instance, structural problems (a high reliance on agriculture and other traditional industries) and problems related to social and institutional failure (as reflected in the levels of long-term unemployment). In general, these counteracting factors appear to exist in combination, indicating that the disadvantages of 'backwardness' are complex and interrelated, and cannot be reduced to a single dimension such as, for instance, the location of a region on a north–south axis.

4 THE IMPACT OF EU REGIONAL SUPPORT

In recent years EU Structural Funds have been reformed and their funding has increased significantly. The purpose has been to make them more effective in reducing the gap between the more and less advanced regions and strengthening economic and social cohesion in the European Community.[16] Following the 1988 reform, the financial resources allocated to these funds have doubled in real terms. The extent to which these funds have contributed to regional convergence in the EU is not addressed directly in our analysis. One reason for not splitting our sample in two, and analysing growth during 1980–87 and

1988–94 separately, is that we believe these periods are too short for explaining long-term growth. In fact, most of the recent empirical growth literature has used longer time periods than we have. Thus to estimate the effects on long-term growth rates, or income levels for that matter, of increased regional support, seems premature at this stage. One would run the risk of mixing a short-run Keynesian demand effect of higher transfers with the long-term supply effect.

It is worth noting, however, that the criteria used by the Structural Funds when allocating resources to different regions are to a large extent reflected in our analysis. The so-called Objective 1 regions are those which are characterized by low levels of GDP per capita, a large agricultural sector, a poorly developed infrastructure and high long-term unemployment. The Objective 2 regions are those affected by industrial decline. Objective 3 regions are those with long-term unemployment problems for people over 25 years of age, while Objective 4 is concerned with the integration of young people into working life. Funds for Objective 5 regions are partly related to agriculture and partly to rural development. Table 6.5 provides a breakdown of the Structural Funds by country and objective.[17]

Table 6.5 Structural Funds by country and objective, 1989–93 (ECU mn)

Country	Obj. 1	Obj. 2	Obj. 3,4	Obj. 5	Total	Percentage of GDP
Belgium	—	214	344	182	740	0.1
Denmark	—	25	171	206	402	0.1
Germany	2955	581	1054	1425	6015	0.1
Greece	7528	—	—	—	7528	2.7
Spain	10171	1506	837	586	13100	0.7
France	957	1225	1442	2283	5907	0.1
Ireland	4460	—	—	—	4460	2.7
Italy	8504	387	903	959	10753	0.3
Luxembourg	—	12	11	32	55	0.2
Netherlands	—	165	405	155	725	0.1
Portugal	8450	—	—	—	8450	3.1
UK	793	2015	1502	506	4816	0.1
EUR12	43818	6130	6669	6334	62951	0.3

Source: EC (1996, Tables 24 and 25).

As can be seen from the table, support from the Structural Funds is small in most EU countries. On average it amounts to 0.3 per cent of GDP. There are only three countries for which such support really matters on a national scale:

Greece, Ireland and Portugal. For these countries the total support from the funds amounts to roughly 3 per cent of annual GDP in each country per year. The regions that receive support in these countries all fall under Objective 1 (the poorest regions). Spain also receives an above average share (0.7 per cent), most of which also falls under Objective 1.

If regional support from the EU leads to significantly higher growth in the supported regions, our analysis may be biased, because we omitted an important factor affecting regional growth. There are two possible effects that such support might have on our analysis. First, in the countries for which such support is widespread, it may show up in nationwide growth, that is, in the estimated country-specific factors. However, by inspection it turns out that the two countries in our sample that have received such support on a substantial scale, Greece and Portugal, both receive a negative contribution from the country-specific factor. Hence, although a positive effect on the national level is a theoretical possibility, it is not something that appears to be reflected in our data.

The other possible effect implies that the estimates obtained on variables that are also Objective 1 criteria may be biased. For instance, our analysis suggests that poor regions may grow rapidly through imitation of more advanced technologies developed elsewhere. But the reality could be that these regions grow fast because they get regional support! If so, we would be likely to overestimate the contribution from the potential for catch-up. Or take agricultural regions, which are also supported. Failing to take into account the support to agricultural regions might lead us to underestimate the negative effect on growth of a large agricultural sector. However, the results obtained from splitting the sample (Tables 6.3–6.4) show that there is no significant difference between the estimates obtained for these two variables in the two subgroups of the split(s). Thus it appears that in these cases the hypothesis of a bias, although possible in theory, is not of great practical relevance. This may be different, though, for long-term unemployment (which is, as mentioned, also one of the Objective 1 criteria) in the split based on that variable. In this case we found a significantly smaller negative effect of unemployment among high unemployment regions, consistent with the assumption of a downward bias in the estimate. Hence it is possible that by failing to take into account various types of support that prop up growth in high unemployment areas,[18] we have underestimated somewhat the negative impact that unemployment has on growth in these regions.

Another way to look at this issue would be to investigate what happened to regional dispersion within countries in the last decades. The assumption, then, would be that countries that have attracted a lot of regional support should be more likely than others to converge internally. Most of the Objective 1 regions, that receive the lion's share of the support, are located in Greece, Ireland,

Portugal, Spain and southern Italy. When we look at the regional dispersion within EU countries (see Figure 6.1), it becomes clear that neither within Greece nor Italy has increased regional support through the Structural Funds prevented regional dispersion from increasing. There is also no change in the trend after 1988 for these countries. For Portugal and Spain, on the other hand, there is some evidence of a long-term decline in dispersion. But for Spain there is no evidence of faster internal convergence after 1988, and for Portugal data problems make it difficult to make an assessment.[19] Thus, there is no strong support in the data for the idea that increased support from the Structural Funds had a great impact on dispersion.

5 CONCLUSIONS

The evidence presented in this chapter shows that the level of aggregation in the empirical analysis matters a lot for our interpretation of growth processes. If one looks at the country level there appears to be a tendency towards long-run convergence in productivity and income levels in the European Union. However, this tendency masks important differences between regions within the same country. In fact, for most countries, there is either little change in regional dispersion, or a tendency towards divergence. This appears to happen in spite of increasing efforts at the EU level in the form of increased contributions to backward regions from the Structural Funds.

How is this to be explained? We looked at these developments from three closely related angles: *technology*, *structure* and *institutions*. The technological perspective is based on the distinction between innovation and diffusion of technology. Regions at the frontier may receive a growth stimulus from investments in innovation activities (R&D), while backward regions may benefit from the adoption of new technologies at a relatively low cost. The evidence considered in this chapter shows that both sources are important for growth. However, many backward regions have failed to take advantage of the potential for diffusion. Lack of data prevents us from exploring many potentially relevant aspects of this, but among the factors that we were able to take into account, the industrial structure of the region and its ability to fight long-term unemployment appear to be among the most important. The evidence shows that, conditional on other factors, the most successful regions are those specialized in services and high-tech industry, while agricultural regions and those specialized in traditional industries are at a disadvantage. This is consistent with the view that technological progress in recent decades, dominated by ICT, has been more beneficial to services than to most other sectors of the economy.

In previous work two of us showed that, conditional on other factors, regions that receive support from the Structural Funds do not grow faster than other regions (Fagerberg and Verspagen, 1996). Here, we have looked further into this issue and found that there is little evidence suggesting that support from the Structural Funds, including the increase in these funds in recent years, has had any significant effect on growth at the regional or national level. The findings in this chapter suggest that to achieve growth regional support schemes should give priority to policies that:

- increase the ability of the regions to absorb new technologies;
- change the industrial structure (away from agriculture and traditional industries), and lead to a reduction in long-term unemployment.

Why existing policies have failed to produce significant growth effects is not dealt with in great detail here, and will be an important topic for further research in this area. One hypothesis might for example be that these policies are not designed in a way that encourages growth, that is, do not sufficiently take into account the points made above. Another possibility would be that any positive effect on growth that these policies might have is counteracted by other policies at the EU or national level, such as, for instance, the Common Agricultural Policy (CAP).

One factor that makes the design of regional policy difficult is that similar policies may have quite different consequences in different environments. This has to do with the fact that regions may differ not only by initial conditions, investments in R&D and physical capital and so on, but also in the dynamics, that is, in the impact that these variables have in the economy. To explore this issue we tested for possible differences in the impact of variables across regional subgroups defined by the three dimensions mentioned above (technological, structural and institutional). We also included a fourth dimension, based on a north–south (geographical) divide. The results showed that a simple north–south dichotomy explains very little. However, what seems to be a fairly robust finding is that there exists a high R&D, high productivity subgroup with its own dynamics. What distinguishes these high R&D regions from the rest is primarily that R&D matters a lot in the former, while it is of little importance (or contributes negatively) in the latter. Also the tendency towards deindustrialization (negative impact of specialization in industry) is much weaker among high R&D regions than for the others, indicating that high R&D industry is less affected by these negative trends than industry at large. These findings may serve as a warning against R&D euphoria in backward regions that lack the necessary infrastructure. Clearly, creating technological capabilities in backward regions demands much more than R&D, and it is an important task for future research to explore the interaction between the various factors that take part in these processes.[20]

NOTES

1. The performance of regions in the US, Japan and Europe is analysed in a series of papers by Barro and Sala-i-Martin (see, among others, Barro and Sala-i-Martin, 1991, 1992; Sala-i-Martin, 1996). For a more comprehensive treatment of the European case, see Neven and Goyotte (1995), Fagerberg and Verspagen (1996) and Fagerberg, Verspagen and Caniëls (1997).
2. Some missing observations in the REGIO database have been collected from other publications or are estimates by the authors.
3. To make this measure comparable over time we divided regional GDP per capita by the mean value in the aggregate we look at (that is, EU9, EU12 or the individual country).
4. Molle (1980), now a classic study in this area, relies on so-called Theil-indices of GDP per capita. The Theil-index weights the regional per capita incomes using a measure of the size of the regions as weights. This implies that dispersion may change only because of changes in relative size and not because of changes in the dispersion of per capita incomes per se. A measure such as the standard deviation of the log of relative GDP per capita, used here and by many others, gives all regions the same weight independent of size. See also Dignan (1996).
5. This result is also borne out in the EC (1996, p. 19) for a somewhat shorter period.
6. Note that the figures for Portugal between 1985 and 1990 reflect adjustments (shifts) in the data for certain regions. The hat-shaped curve that these adjustments give rise to is clearly a statistical artifact. What is of interest is the long-run trend.
7. For the Netherlands, the changes in oil (and gas) prices have influenced the incomes of Groningen so that the dispersion dropped dramatically from 1985 to 1987.
8. Italy stands out with the largest within-country dispersion, reflecting Italy's well-known north–south divide. For the other eight countries the internal dispersion as measured by the standard deviation of log per capita GDP (in PPP relative to the average) of each country is now between 0.15 and 0.20 (1995). Hence, apart from Italy, there has been a convergence towards a common level of dispersion.
9. The hypothesis that technological catch-up requires substantial efforts and capabilities in the receiving country is discussed and tested in Fagerberg (1987, 1988). Verspagen (1991) and Amable (1993) analyse the possibility that countries without the necessary assets may end up in a low-growth trap. For an overview of empirical work on catch-up and growth, including its theoretical underpinnings, see Fagerberg (1994).
10. All data for the variables described below are taken from the EUROSTAT REGIO database.
11. Industry as used here includes fuel and power, manufacturing and construction. The remaining part of total employment when agriculture and industry are deducted is services, which therefore cannot be included as a separate variable.
12. Reverse causation cannot be excluded, of course, but previous research shows that the distribution of unemployment is roughly constant through time. Hence the high unemployment regions at the beginning of the period also had high unemployment at the end, and vice versa (Fagerberg, Verspagen and Caniëls, 1997). This seems to point to persistent social, institutional and economic problems in these regions that depress growth.
13. In the paper by Durlauf and Johnson (1992), and the article by Fagerberg and Verspagen (1996), the samples of j and $n-j$ observations are subsequently subjected to the same splitting procedure repeatedly, until the number of degrees of freedom becomes too small to allow a new split, thus building so-called regression trees. In this chapter, we limit ourselves to just one split.
14. The variable was defined as the most southbound longitude that crosses a region.

15. It is possible that our measure of technological capability, R&D employment, is biased towards large firms in high R&D industries and countries, and fails to reflect the efforts by small firms, firms engaged in imitation, and industries in which learning – rather than organized R&D – dominates technological progress. This is an area where further research seems not only fruitful but absolutely necessary. For an analysis of differences in modes of technological accumulation across industries, see Pavitt (1984, 1988).
16. For an analysis of regional policy in the EU, including its rationale and the need for reform, see Begg and Mayes (1993) and Begg (1997).
17. Note that funds disbursed to regions according to Objective 3 and 4 criteria are not regional specific in nature. But they may still have an allocation biased towards peripheral regions as high long-term unemployment and low rates of labour participation are quite typical features of such regions.
18. Note also that the high unemployment regions grow faster than any other subgroup included in Table 6.3.
19. As mentioned the figures for Portugal between 1985 and 1990 reflect adjustments (shifts) in the data for certain regions. The hat-shaped curve that these adjustments give lead to is clearly a statistical artefact.
20. See also Chapter 8 by Dalum et al. in this volume.

REFERENCES

Abramovitz, M. (1994), 'The Origins of the Postwar Catch-Up and Convergence Boom', in J. Fagerberg, B. Verspagen and N. von Tunzelmann (eds), *The Dynamics Of Technology, Trade and Growth*, Aldershot, UK: Edward Elgar, pp. 21–52.

Amable, B. (1993), 'Catch-Up and Convergence: A Model of Cumulative Growth', *International Review of Applied Economics*, **7**, 1–25.

Barro, R. J. and X. Sala-i-Martin (1991), 'Convergence across States and Regions', *Brookings Pap. Econ. Act.*, **1**, 107–82.

Barro, R.J. and X. Sala-i-Martin (1992), 'Regional Growth and Migration: A Japan–United States Comparison', *Journal of the Japanese and International Economics*, **6**, 312–46.

Begg, I. and D. Mayes (1993), 'Cohesion in the European Community. A Key Imperative for the 1990s?', *Regional Science and Urban Economics*, **23**, 427–48.

Begg, I. (1997) 'Reform of the Structural Funds after 1999', *European Planning Review*, **5**, 675–89.

Dignan, T. (1996), 'Regional Disparities and Regional Policy in the European Union', *Oxford Review of Economic Policy*, **11** (2), 64–95.

Durlauf, S.N. and P.A. Johnson (1992), 'Local versus Global Convergence: Across National Economies.' *NBER Working Paper*, No. 3996. Cambridge, MA: National Bureau of Economic Research.

European Commission (1996), 'First Report on Economic and Social Cohesion', Brussels: European Commission.

Fagerberg, J. (1987), 'A Technology Gap Approach to Why Growth Rates Differ', *Research Policy*, **16**, 87–99.

Fagerberg, J. (1988), 'Why Growth Rates Differ', in Giovanni Dosi et al. (eds), *Technical Change and Economic Theory*, London: Pinter, pp. 432–57.

Fagerberg, J. (1994), 'Technology and International Differences in Growth Rates', *Journal of Economic Literature*, **32**, 1147–75.

Fagerberg, J. and B. Verspagen (1996), 'Heading for Divergence? Regional Growth in Europe Reconsidered', *Journal of Common Market Studies*, **34**, 431–48.

Fagerberg, J., B. Verspagen and M. Caniëls (1997), 'Technology, Growth and Unemployment across European Regions', *Regional Studies*, **31**, 457–66.
Kaldor, N. (1967), *Strategic Factors in Economic Development*, Ithaca, NY: Cornell University.
Molle, W. (1980), *Regional Disparity and Economic Development in the European Community*, Westmead: Saxon House.
Molle, W. and R. Cappellin (eds) (1988), *Regional Impacts of Community Policies in Europe*, Aldershot, UK: Edward Elgar.
Neven, D. and C. Goyette (1995), 'Regional Convergence in the European Community', *Journal of Common Market Studies*, **33**, 47–65.
Pavitt, K. (1984), 'Patterns of Technical Change: Towards a Taxonomy and a Theory', *Research Policy*, **13**, 343–74.
Pavitt, K. (1988), 'International Patterns of Technological Accumulation', in N. Hood and J.E. Vahlne (eds), *Strategies in Global Competition*, London: Croom Helm.
Sala-i-Martin, X. (1996), 'Regional Cohesion: Evidence and Theories of Regional Growth and Convergence', *European Economic Review*, **40**, 325–52.
Solow, R.M. (1956), 'A Contribution to the Theory of Economic Growth', *Quarterly Journal of Economics*, **70**, 65–94.
Verspagen, B. (1991), 'A New Empirical Approach to Catching Up or Falling Behind', *Structural Change and Economic Dynamics*, **2**, 359–80.

7. Europe: one or several systems of innovation? An analysis based on patent citations*

Per Botolf Maurseth and Bart Verspagen

1 INTRODUCTION

European integration has traditionally been aimed at reduction of barriers to intra-European trade and factor mobility. This has been achieved by the abolition of tariffs and import duties, by liberalization of capital movements and reduced barriers to foreign direct investments, by legislation facilitating mobility of people across the European Union, and by the abolition of various so-called non-tariff barriers to trade under the 1992 programme.

Has a similar degree of integration been reached in the field of technology and innovation? This is an important question since the extent to which a nation or region can assure access to external technological knowledge may be decisive for its economic performance.

The literature on so-called national systems of innovation (for instance, Lundvall, 1992) focuses on the ways in which knowledge flow takes place. Knowledge flows and the factors which have an impact on them are not easy to quantify, however. In the national systems of innovation approach, various factors related to institutions, culture and history are identified as important. From another perspective, Coe and Helpman (1995) have pointed to trade as an important channel for diffusion of technology embodied in physical goods. Diffusion of disembodied technology, in the form of general or specific knowledge, is also believed to be of importance. Although knowledge flows are to some extent related to trade flows, the general tendency to liberalize trade flows in the European Union does not necessarily imply a proportionate increase of knowledge flows.

Much of the recent literature argues that technology spillovers are to an important extent local (Morgan, 1997; Jaffe, Trajtenberg and Henderson, 1993). The reason for this is that, despite modern communications techniques, due to the tacitness of knowledge, frequent face-to-face contact, or mobility of knowledge workers are still important channels of knowledge spillovers. In this chapter, we use patent citations as an indicator of disembodied technology flows. Such citations may be interpreted as external knowledge flows or technological spillovers. In the analysis presented here, we will focus on the impact of geography, national borders, economic development and technological specialization on the flow of technological spillovers between a set of European regions.

The rest of this chapter is organized as follows. Section 2 discusses some implications of spillovers in general and localized spillovers in particular for economic development at the country and regional levels. Section 3 describes innovative capability in European regions with use of data on patenting. Section 4 provides a descriptive analysis of regional technological interaction as evidenced by patent citations. In section 5 econometric evidence on the determinants of technological spillovers between European regions is presented. The concluding section 6 summarizes the empirical findings and points out some directions for future research. Data construction and data sources are discussed in Appendix 7.1.

2 ECONOMIC GROWTH, REGIONAL DEVELOPMENT AND TECHNOLOGICAL SPILLOVERS

An important characteristic of technological knowledge is that it can be used without being exhausted. Technology is also cumulative in nature, because it is based on previously gained insights. Furthermore, technological knowledge is seldom (completely) limited to the person or firm that developed it and has the property rights to it. In other words, technological spillovers take place. In the recent formal growth models in the neoclassical tradition, increasing returns through spillovers make endogenous growth possible. Without such spillovers, economic growth either ceases in the long run (see Grossman and Helpman, 1991, Chapter 3), or is 'explained' as a completely exogenous process (as in the old neoclassical model of the 1950s).

When spillovers are industry-specific, specialization in certain industries may result in higher growth than specialization in other industries, and the specialization pattern of a country or region is then likely to have an impact on economic growth. If spillovers are geographically concentrated, knowledge stocks may accumulate in proportion to local industrial activity. Thus increasing returns resulting from spillovers may be bounded within geographical limits. Localized spillovers thereby facilitate clustering of economic activity.[1] To reap the benefits of local spillovers, production is established near to pre-existing production. External effects from neighbouring establishments increase profitability of further establishments.

Alfred Marshall observed early on that knowledge spillovers may play a crucial role for clustering of economic activity.[2] In addition to obvious explanations such as endowments of natural resources, Marshall referred to technological spillovers as one of three possible explanations[3] for clustering of economic activity (Marshall, 1948, p. 271):

> When an industry has thus chosen a locality for itself, it is likely to stay there long: so great are the advantages which people following the same skilled trade get from near neighbourhood to one another ... [I]f one man starts a new idea, it is taken up by others and combined with suggestions of their own; and thus it becomes the source of further ideas.

The importance of geography for diffusion of knowledge was also recognized by Raymond Vernon as a basis for his product cycle theory.[4] Vernon showed how localized knowledge and technological opportunities might envisage introduction and production of new products in advanced markets.

Kaldor (1978, p. 143), reflecting on uneven regional development, analysed the role of localized dynamic increasing returns as a result of, among other factors, 'the opportunities for easy communication of ideas and know-how'. Kaldor, inspired by Allyn Young's theorizing on technological spillovers as a source of aggregate increasing returns (Young, 1928), hypothesized that regional development was subject to a principle of 'circular and cumulative causation' in which regional economic progress (or stagnation) is the seed of further progress (or stagnation). Thus, uneven regional development may be an inherent outcome of decentralized economic processes in the absence of counteracting economic policy. Kaldor pointed out that such processes of cumulative causation made a case for regional policies.

Previous empirical research has established that geography may indeed be important for technology spillovers. Analysing patent citations, Jaffe, Trajtenberg and Henderson (1993) found intra-national citations (national patents citing national patents) and intra-state citations (citations to patents originating in the same state) to occur more often than expected from the distribution of patenting activity, using US patent statistics. Similar results were obtained in Jaffe and Trajtenberg (1996) where it also was found that the geographical concentration of spillovers decreased over time. In the case of Sweden, Sjöholm (1996 and 1997) found citations to patents in neighbouring countries to occur more often than to patents originating in more distant countries. Whether these findings extend to Europe as a whole is one of the major issues to be addressed in this chapter.

There are also factors that can be identified as stimulating the flow of knowledge. The so-called technology gap theory on economic growth and international trade deals with the (international) diffusion of technological knowledge (Fagerberg, 1994). This theory focuses on how countries ranking low on the productivity ladder may catch up with leading countries. Diffusion of technology facilitates the potential for catch-up, but technological progress on the frontier increases the height of the ladder to climb.[5] The ability to adapt new technologies depends on institutional infrastructure, education, geography and resources devoted to R&D. The technology gap theory has increased the understanding of critical factors of catch-up with the technological leading countries (for instance, finance, the educational system and politics, see, *inter alia*, Abramovitz, 1985; Fagerberg, 1988; Verspagen, 1991). Fagerberg (1994) concludes a survey on the literature on the catch-up debate with the following: 'Indeed, what the whole literature, from Gerschenkron onwards, suggests is that catching up is very difficult, and that only countries with appropriate economic and institutional characteristics will succeed.'

It thus appears that the absorptive capability of a country or region is crucial for clustering. The cumulative nature of technology and the localness of spillovers bring with them a tendency for clustering, and the extent to which this tendency will be counteracted by wider technology diffusion depends on absorptive capacity. If there are large differences in terms of absorptive capacity, a considerable degree of clustering may arise (depending on whether the peripheral regions have high or low absorptive capacity), whereas if all regions have high absorption capability, spillovers flow easily, and the spread of economic activity will be more even.

An important question, in an increasingly integrated Europe, is the extent to which national systems of innovation are still relevant. Increased integration indicates less importance of national borders. In Europe, economic integration is combined with supranational institution building to support regional development (structural funds), exchange of students, cooperation between universities and R&D laboratories and infrastructure projects. This process raises the question whether a European system of innovation will come to supplement the national systems.

On the other hand, studies of national systems of innovation highlight important path-dependent aspects of such systems. One example is technological spillovers that are somewhat specific in scope (sectorally, geographically, etc.). Reduced importance of national borders and policies may imply that systems of innovation become less national but still geographically concentrated. Thus analyses of the innovative capacity in Europe should incorporate both distinct European aspects (for instance in terms of a European system of innovation) and variety at national and regional levels (national and regional systems).

The system of innovation, and whether it can be characterized as European, national or regional, thus provides a crucial link between localized spillovers (which lead to clustering) and diffusion of technological knowledge (leading to convergence). Our analysis in the next sections will be aimed at answering the question of how knowledge flows in the European innovation system. Can we still observe, despite increased integration since the 1950s, factors that hinder the flow of knowledge through the system? Do we see one truly European system of innovation, in which knowledge spillovers flow between all relevant units (for instance, regions), or do we have instead many isolated innovation systems that only interact marginally with each other?

3 TECHNOLOGICAL COMPETENCIES IN EUROPEAN REGIONS

As is well known from evidence at the country level, there are large differences between European countries in terms of technological competencies (see Chapter 1). In terms of R&D intensity (R&D expenditures as a percentage of GDP), large differences between European countries exist. From the analysis of differences in GDP per capita in the European Union (see Chapter 6 by Cappelen, Fagerberg and Verspagen in this volume), we know that regional differences in GDP per capita are much larger than across countries. Because there is a close correlation between technological competencies and GDP per capita (see Fagerberg, Verspagen and Caniëls, 1997), one might expect that regional differences in terms of technology in the European Union are large.

It is the aim of this section to investigate the extent of these differences. Patent statistics will be used to this end. Such statistics are often used as an indicator of technological strength of a country or region.[6] The fact that patent statistics are output indicators rather than input indicators has some advantages as well as disadvantages. The main advantage is that one is able to circumvent the issue of R&D productivity ('the number of innovations per unit of R&D'), and that patent statistics are available for a wider set of regions and a longer time period than R&D statistics. The main disadvantages lie in the problem that simple patent counts do not take into account differences in the quality of innovations, that many patents do not lead to innovations, and that the propensities to patent may differ between sectors. Despite these differences, patent statistics are widely used to analyse regional differences in innovation in the European Union (for instance, Caniëls, 1996; Paci and Usai, 1997; Verspagen, 1997a and 1997b).

The expectation of a correlation between GDP per capita and patenting between European regions is indeed confirmed by the data. The rank correlation between GDP per capita in 1994 and the share in patent applications at the European Patent Office (EPO) over the period 1979–96 is 0.67.

Figure 7.1 gives an overview of patenting activity in European regions. The map gives four groups (quartiles) of regions, based on the number of patent applications at EPO over 1979–96. All applications are assigned to the region of the home address of the inventor, so we rule out any bias resulting from the fact that patents are often applied for from a location other than the one where the invention was made.[7] The ranking of the regions by number of patent applications is indicated in the figure.

Germany comes out with the highest activity. All regions in the former West Germany rank in the highest group. Even the eastern part, however, has high values, with all regions except one ranking in either the first or second group. This may partly have to do with the fact that the main office of EPO is located in Munich, but it is unlikely that this has a strong impact. Patents can be filed in any language, and all countries have patent lawyers fully qualified to handle

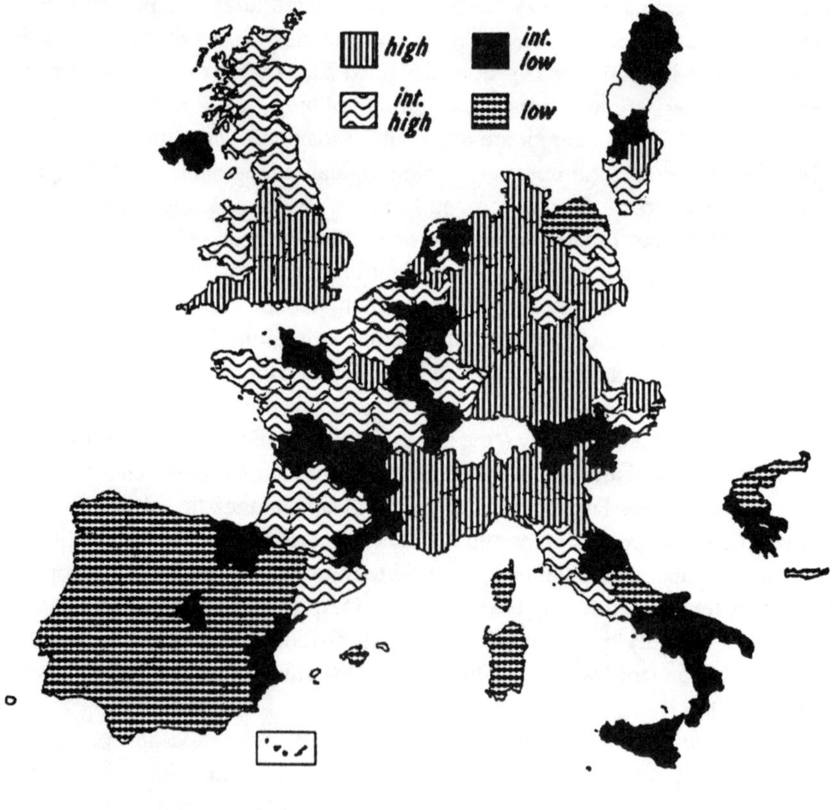

Figure 7.1 Patent applications at EPO 1979–96 (share in total)

EPO applications. We thus interpret the German result as confirming the technological leadership of this country in the European context.

The other members of the high patenting activity group are spread out over six countries: the United Kingdom, France, Italy, Austria and Sweden. It is noteworthy that there is a clear amount of clustering in two areas: north Italy combined with southeast France, and England. Also in the Netherlands, patenting activity clusters in two adjacent regions. In Belgium, France, Austria and Sweden, the regions with capital cities rank high.

Note that in general, patenting activity in the south of Europe is lower than in the north, as could be expected on the basis of GDP per capita data. Portugal, Spain and Greece are the only countries without a region ranking in the highest activity quartile, and in Italy the high activity regions are located in the north. In fact, in the set of regions consisting of Portugal, Spain, Greece and south Italy, there are only two regions which rank in the 'intermediate high' quartile (regions around Rome and Barcelona); all other regions rank lower. In Portugal, all regions rank in the lowest quartile,

Figure 7.2 adjusts the raw patent applications data for the size of the region by dividing the number of patents used in Figure 7.1 by the population of the region in 1990. This mainly has the effect of reducing the dominance of the larger countries such as Italy, the UK and Germany, in favour of smaller countries such as Austria and the Netherlands. Austria now ranks almost fully in the highest quartile, with only two regions ranking in the second group. In the Netherlands, four regions rank in the top group, as do two in Sweden. The large clusters of high activity in north Italy and England are reduced in size, although most of the regions in these clusters still rank in the 'intermediate high' group.

The division between north and south still remains clearly visible. Portugal still ranks completely in the lowest group, as does Greece, and the largest part of Spain (three regions in this country rank in 'intermediate low', the rest in 'low'). South Italy also ranks very low. In fact, no region north of the Pyrenees ranks in the 'low' group.[8]

The conclusion on the geographical spread of inventive or innovative activity over Europe is thus that there is a fair amount of concentration. This concentration occurs in various dimensions. It is perhaps most visible in the north–south context, where we find the familiar pattern of high innovation activity in the north, and low activity in the south. However, also in the within-country dimension, concentration occurs. Each of the countries in our maps clearly shows some geographical concentration of patenting.

The degree to which patenting is geographically concentrated differs, however, between industries. In order to investigate this dimension, we assigned each of the patent applications to one or more of 22 manufacturing

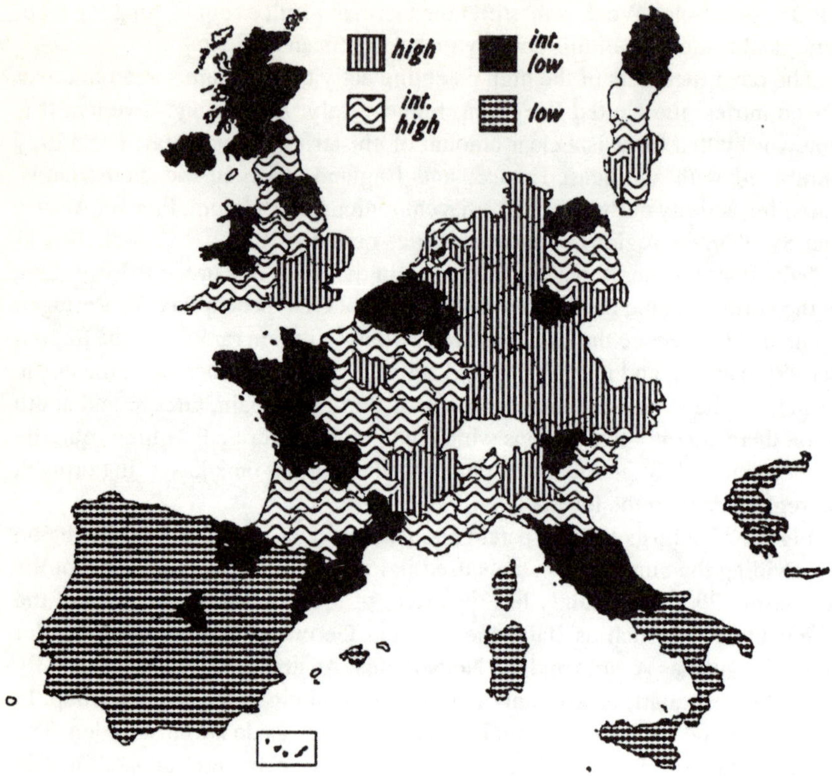

Figure 7.2 Patent applications at EPO 1979–96 per head of population in 1990

industries, according to the MERIT concordance table between IPC and ISIC (Verspagen, van Moergastel and Slabbers, 1994). We thus have, for each region, not only the total number of patent applications, but also the spread of these over 22 sectors.

It turns out that for each sector, particularly the so-called high-tech ones,[9] the (rank) correlation between the total amount of patents in the regions and the sector-wise number of patents by region is quite high. It would thus be redundant to repeat the maps shown for total patenting at the sectoral level. Instead, we present, in Table 7.1, a more synthetic measure of concentration, in the form of the Herfindahl index. This index is defined as the sum (over regions) of squares of regional shares of patenting.[10] A high number indicates high concentration.

Table 7.1 presents the results of these calculations for the 22 sectors. The two sectors with the highest geographical concentration are both high-tech sectors, that is, pharmaceuticals and computers. Other high-tech sectors also show high concentration (for instance, aerospace, electronics). The other sectors for which patenting is relatively concentrated are all scale-intensive sectors: ferrous basic metals, chemicals, electrical machinery, refined oil, ships and boats. The high geographical concentration of patenting in these sectors may thus well be the result of a high concentration of economic activity over space, rather than the result of some inherent tendency for technological activity to be clustered.

Table 7.1 Concentration of patenting over European regions for 22 manufacturing sectors; total patenting and high-tech patenting

	HF-index
Pharmaceuticals	0.075
Computers and office machines	0.073
Ferrous basic metals	0.071
Electrical machinery	0.068
Aerospace	0.067
Chemicals	0.064
Refined oil	0.059
Ships and boats	0.056
Electronics	0.055
Motor vehicles	0.051
High-tech aggregate	0.049
Other transport	0.048
Paper and printing	0.047
Textiles, apparel, leather	0.046
Instruments	0.046
Non-ferrous basic metals	0.045
All sectors aggregate	0.043
Machinery	0.042
Metal products	0.042
Wood and wooden products	0.041
Food, drinks and tobacco	0.040
Glass, stone and clay	0.039
Rubber and plastic products	0.035
Other manufacturing	0.034

Source: Calculations based on EPO data.

4 TECHNOLOGY SPILLOVERS BETWEEN EUROPEAN REGIONS

From the point of view of a European innovation *system*, what matters is not only the distribution of activities over regions, but also the way in which regions 'interact' with respect to technology. As pointed out in section 2, technology spillovers may have important effects on economic development in and across regions and countries.

Griliches (1979) distinguishes between two types of spillovers, that is, rent spillovers and pure knowledge spillovers. Rent spillovers are *pecuniary* spillovers which result when the innovating firms are unable to raise prices proportionally to the quality improvements of their products. For the firms that use these products as inputs, this results in a better quality–price ratio, which is interpreted as a spillover.[11]

The concept of pure knowledge spillovers, on the other hand, is related to 'the impact of the discovered ideas or compounds on the productivity of the research endeavours of others' (Griliches, 1992, pp. 30–1). This corresponds to the impact of 'general knowledge' on the productivity of R&D in Romer's (1990) model. In this context, one may think of 'imitative' spillovers (that is, one firm copying an innovation by another firm), or 'idea-creating' spillovers (when an innovation leads to an idea for another innovation).

Because the impact of rent spillovers is largely related to traded inputs, one could expect that the importance of such spillovers increases when trade barriers in Europe are reduced. This does not necessarily hold for pure knowledge spillovers. Unfortunately, pure knowledge spillovers are a difficult concept to operationalize, given the available indicators. Because no data are available on the R&D-financing links between regions or countries, we cannot follow the more traditional methodology to use these data as an indicator of interaction.[12]

Instead we use data on patent citations as an indicator of interaction. This follows earlier contributions (using data at the national level) in Verspagen (1997a, 1997b), building on a method proposed by Jaffe, Trajtenberg and Henderson (1993).

Each patent application must refer to previous patent applications. The purpose of patent references is to preclude double patenting of innovations and eventually to limit patent protection. Also, patent references indicate relevant established knowledge for new innovations. One may thus straightforwardly interpret such patent references as indicators of spillovers of knowledge from the cited patent to the citing patent. For the purpose of this chapter, citations in European patents are used as our measure of knowledge spillovers. However, it has to be kept in mind that the large majority of the patent citations is added by the EPO patent examiners, which implies that the inventors may not have been aware of the cited patent. Still, the citation link may be seen as an indicator of technological relevance. This certainly indicates potential spillovers, although this potential may not have been realized in all cases. However, patents are public knowledge, so professional R&D laboratories can to a certain extent be assumed to be able to extract useful knowledge from existing patents.

It should be emphasized that the concept of knowledge spillovers is much broader than what is captured by our indicator. In terms of the distinction by Griliches introduced above, we look at a specific form of pure knowledge spillovers. Rent spillovers are left out of the analysis completely. Even within the category of pure knowledge spillovers, however, patent citations are only a part of the complete story. In order for patent citations to take place, both the spillover-receiving and spillover-generating region must be actively engaged in R&D, leading to patent applications.

Thus our analysis will refer to a specific form of knowledge spillovers. This may seem to be a narrow perspective, but has the advantage that we can make use of a very detailed and precise database, in contrast to the more general indicators of spillovers that have been used in other parts of the literature.

The citation data were used to set up a list of pairs of cited and citing patent applications.[13] The regional citation list consists of pairs of 112 European regions, plus seven country aggregates (so the total number of observations is 119 × 119). Four of these countries are European ones for which no regional breakdown of the data is available. The remaining three are the USA, Japan and 'other countries' (other than the USA, Japan or European countries).

The region-by-region list has a relatively high concentration of intra-regional citations. Approximately 35 per cent of all citations are for observations when the cited and the citing region are one and the same (note that only 0.84 per cent of all observations in a 119 × 119 list are for the cases when the citing and the cited region are identical). The data do not enable us to make a distinction between citations within a firm or establishment, and citations between two different firms located in the same region. Obviously, this distinction is of great relevance for a discussion on regional innovation systems. The distinction between Silicon Valley and Route 128 is illustrative in this respect (see Saxenian, 1994). In Route 128, one finds large, vertically integrated firms which can be expected to have a large number of intra-firm citations. In Silicon Valley one finds many more small firms, operating in an open and interactive system, with, expectedly, more interfirm (but intra-regional) citations. Saxenian (1994) argues that the two innovation 'systems' can be expected to show quite different levels of performance (the Silicon Valley system is argued to be more efficient).

With the data we have, however, we are unable to separate the Route 128s from the Silicon Valleys. This is the reason why we will generally leave out the intra-regional citations from the analysis. This does not imply that intra-regional and intra-firm citations are not an interesting phenomenon. It merely indicates the limitations of the data.

The inter-regional citations show a highly skewed distribution. This is shown in Figure 7.3. Slightly more than half of all regional pairs (7609 pairs of regions) never cites each other's patents. The frequency of citations gradually declines for more intensive citation links. There are only 71 pairs of regions for which the number of citations is 200 or more. Note that the total number of inter-regional citations is approximately 110 000. The average number of citations for all pairs of regions is 8.6 and the average for those that cite each other is about 17.

From this we conclude that strong technological spillovers between European regions, as far as they are related to patent citations, are only found between a relatively small number of regions. These are also the regions which are relatively active in terms of patenting (see Figures 7.1 and 7.2).

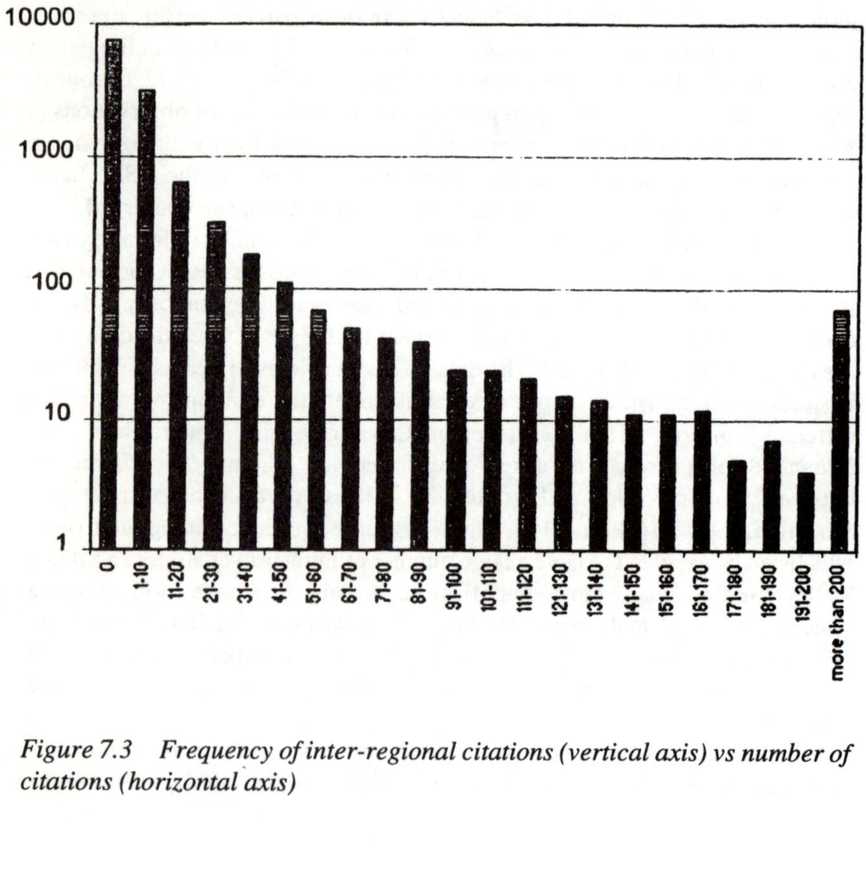

Figure 7.3 Frequency of inter-regional citations (vertical axis) vs number of citations (horizontal axis)

5 SPILLOVERS BETWEEN EUROPEAN REGIONS – ECONOMETRIC RESULTS

The aim of this section is to systematically investigate the pattern of patent citations in Europe based on the data just described. Of particular interest is the extent to which geography affects the technological interaction between European regions. The previous sections, however, have indicated several other factors of potential importance for knowledge flows. Thus the effects of technological specialization, productivity gaps (between spillover-receiving and spillover-producing regions), innovative activity, and of national systems of innovation are also taken into account. The analysis is confined to the European regions only (minus Guernsey and Isle of Man), as comparable data on all variables are not available on overseas regions. The new data set consist of 112 × 112 observations (including the four European countries for which regional breakdowns are not available).

To explore whether spillovers are influenced by geography and country borders, we have to control for other factors which are also likely to matter for spillovers. For example, if patent seekers from Ludwigshaven in Germany (the headquarters of the chemistry giant BASF) cite other German patents, this may be due to several distinct effects which do not all necessarily reflect the impact agglomeration.

First, as pointed out in section 3 above, German regions are highly innovative as compared to the European average. Thus German patents are more likely to cite each other just because of the above-average German patent activity. This effect should not be taken as evidence of clustering effects of spillovers. Technologically active regions are a priori supposed to cite each other more often than technologically inactive regions. In fact, the data reveal a high correlation between citations and the numbers of patents in the related regions.[14] This is largely due to the fact that patenting in both the cited and the citing region is a necessary condition for any reference between them at all. For this reason, we construct a dependent variable where the numbers of citations between two regions are expressed as a fraction of the sum of patents in the citing and the cited regions. In other words, what we are trying to explain is not the absolute amount of spillovers as indicated by patent citations, but rather the intensity of this flow compared to total patenting activity in the regions.

Second, patent applicants affiliated to, say, BASF are a priori more likely to cite patents related to chemistry than patents related to other sectors. If Germany is specialized in chemistry, patent applicants from Ludwigshaven are more likely to cite other German patents because of the pre-existing pattern of economic specialization. This type of spillover may be taken as evidence of industry-specific spillovers. As discussed in section 2, industry-specific spillovers may have important implications for the effects of economic specialization.

To take industrial specialization into account, we construct a variable called the 'compatibility index', which makes use of the observed pattern of citations between sectors and the regions' sectoral specialization in patenting. If two regions are specialized in sectors that are often observed to cite each other, this combination of regions receives a high score on the compatibility index. The technicalities concerning this index are discussed in Appendix 7.1. The compatibility index, denoted by s_{ij}, ranges between minus one and one, and the impact of the index on the spillovers between two regions is expected to be positive.[15]

Distance data were constructed by counting the number of regional borders one has to cross to reach one region from another. This yields a list of distances between all pairs of European regions. The distance variable is denoted by d_{ij}. Technicalities regarding the distance data are discussed in Appendix 7.1.

The literature on national systems of innovation seeks to explore how differences in national history, institutions, policy and traditions may affect countries' innovative capability and competencies. This chapter does not aim to explore all aspects of such national systems. To take into account possible effects of national systems of innovation, however, we include a dummy variable for intra-country citations, as well as dummy variables for each cited and citing country in the sample.

Technology gap models point to a potential for poor countries to catch up with economic and technological leaders. However, as noted in section 2, spillovers do not only depend on technology gaps, but also on absorption capability in the lagging country or region, and technological congruence. Taking into account these two variables, Verspagen (1991) argued that while a large gap indicates a large potential for spillovers, it may also imply a low capacity to assimilate spillovers. Thus (very) low-income regions may become stuck in a kind of underdevelopment trap, unable to make use of spillovers from advanced countries. Medium-income countries may be better placed to take advantage of knowledge created in other countries. In other words, the amount of realized spillovers may well be a non-monotonic function of the size of the gap.

It is hard to judge on a priori grounds what constitutes a large gap (that is, one that hinders spillovers more than it creates potential for it) or a small gap (that is, one that stimulates spillovers). Fagerberg and Verspagen (1996), Paci (1997) and Chapter 6 by Cappelen, Fagerberg and Verspagen in this volume indicate that the productivity gaps between some European regions are substantial, and that the disparity is substantially larger at the regional level than at the national level. Thus one would want to allow for a positive as well as a negative impact of the productivity gap on technology spillovers. We therefore allow for a non-linear relationship between spillovers and the productivity gap, by including a GAP-variable, as well as its squared value. The productivity gap variable *GAP* is defined as the log of the ratio of GDP per capita in the spillover-receiving and the spillover-generating region.

Finally, we take into account the possibility that the distribution of patents between the receiving and the generating region may have an impact on the intensity of patent citations. In order to quantify this, we include (the log of) the two regions' share of their total patenting, $P_i/(P_i + P_j)$ and $P_j/(P_i + P_j)$, as explanatory variables. If these two variables receive an equal coefficient in the regressions, this indicates that an equal distribution of total patenting (the spillover-receiving region patents as much as the spillover-generating region) is most conducive for growth. Should any of the two variables receive a higher coefficient than the other, this indicates that a distribution in favour of that particular region is most conducive to spillovers.

We arrive at the following regression model for our indicator of spillovers between regions, SP_{ij}:

$$\begin{aligned} SP_{ij} = \ln\left(\frac{C_{ij}}{P_i + P_j}\right) &= \alpha_0 + \alpha_1 \ln\left(\frac{P_i}{P_i + P_j}\right) + \alpha_2 \ln\left(\frac{P_j}{P_i + P_j}\right) + \alpha_3 \ln d_{ij} \\ &+ \alpha_4 COUNT + \alpha_5 \ln GAP_{ij} + \alpha_6 \left(\ln GAP_{ij}\right)^2 + \alpha_7 s_{ij} \\ &+ \sum_{n=1}^{14} \alpha_{8n} CitedCOUNTRY + \sum_{m=1}^{14} \alpha_{9m} CitingCOUNTRY + \varepsilon \end{aligned}$$

COUNT is a dummy variable for intra-country spillovers. *CitingCOUNTRY* and *CitedCOUNTRY* are dummies for the citing and cited individual countries (14 of each). A significant *CitingCOUNTRY* result indicates that the country's capacity to absorb spillovers differs significantly from the average. A significant *CitedCOUNTRY*-variable indicates country-specific effects in terms of producing spillovers. ε is the error term in the regression. The model is estimated by heteroscedasticity consistent least squares.

The results are reported in Table 7.2.[16] *P*-values are given in parentheses. The general impression is that the model fits the data well. The relatively high R^2 indicates an overall good fit, and most of the coefficients for the structural variables are significant at better than 1 per cent probability level.

Table 7.2 Regression results on spillovers, least squares regression, excluding observations with zero citations. Heteroscedasticity-consistent P-values

Dependent variable: $\ln(C_{ij}/[P_i + P_j])$	$R^2 = 0.47$	Number of observations 6341
Variable	*Coeffcient*	*P-value*
$\ln(P_i/[P_i + P_j])$	0.5174105	0.000
$\ln(P_j/[P_i + P_j])$	0.4984828	0.000
$\ln d_{ij}$	–0.3693099	0.000
COUNT	0.441461	0.000
s_{ij}	0.758228	0.000
$\ln GAP_{ij}$	0.0657392	0.084
$(\ln GAPij)^2$	–0.2888199	0.000

Note: 28 country-specific dummy variables included in regression, but not reported.

The results indicate that there are important barriers to technology spillovers in Europe. This is seen by several results. First, spillovers between a pair of regions decrease significantly with the distance between them. Even if the dependent variable is weighted, making the interpretation harder, the magnitude of the coefficient of the distance variable is large. A 1 per cent increase in distance decreases the expected spillovers (in terms of the constructed weighted dependent variable) between two regions by 0.37 per cent. As an example of the impact of distance, consider the spillovers between Paris and its neighbouring region Picardie. Compared to Picardie, a region that is at distance 2 from Paris (for instance, Wallonie in Belgium) receives 22 per cent less spillovers (*ceteris paribus*). For a region at distance 8 from Paris (such as Sicily), the percentage is 53.[17]

Second, the intra-country dummy variable (*COUNT*) is positive and significant. The magnitude of this variable indicates that country borders significantly hinder knowledge spillovers. This finding gives some support for the importance of national systems of innovation, although we have not investigated whether this result is due to language, institutions or other factors.[18]

Third, the impact of the GAP-variable should be noted. The sign of the estimated coefficients indicates a hill-shaped parabola, that is, knowledge spillovers decrease with the size of the gap (on both sides of the vertical axis). The top of the parabola (that is, the value of the GAP-variable that, *ceteris paribus*, maximizes the amount of spillovers) occurs for a value of ln*GAP* slightly less than zero.[19] This indicates that a small productivity gap (that is, the spillover-receiving region lags somewhat behind relative to the spillover-generating region) is most conducive for spillovers. Very poor regions and very rich regions do not receive many spillovers from other regions. For a very rich region, this may indicate that its high technological competency relative to the other region reduces the potential for learning. For very poor regions, the result indicates that poor regions lack absorptive capacity to benefit from technology developed elsewhere. This result thus gives support for the existence of low-growth underdevelopment traps, as found, for instance, in Verspagen (1991) for a large sample of countries at different levels of development. It also indicates that spillovers do not flow so easily between core and periphery, but rather tend to stay within a group of already relatively well-developed regions.

The importance of technological compatibility for spillovers is also supported. Regions specialized in sectors that are observed to cite each other often, do in fact cite each other more often than average regions do.

Finally, patenting activity in both the cited and the citing region impacts positively, (almost) symmetrically and very significantly on spillovers between each pair of regions. This indicates that regions that patent in approximately equal amounts share most spillovers.

Because of the logarithmic relation used, all observations in which either the number of patents or the number of citations is zero were not included in the regressions. About half of all pairs of regions do not cite each other. Thus, excluding these leaves a biased and not representative sample. The easiest way to include the zero observations is to add a small value to the (raw) citations variable. This makes it possible to take the log of these numbers, although they receive small weights due to the fact that logarithmic functions approach minus infinity for numbers approaching zero. The results reported in Table 7.3 are obtained by adding 0.0001 to the observations of inter-regional patent citations.

The signs of the coefficients are the same as in Table 7.2. However, the magnitudes of the coefficients (except for the compatibility index) are larger than in the case when the zero observations were excluded. Also the fit is somewhat reduced and the linear GAP-variable loses significance. These changes relative to Table 7.2 may partly reflect the fact that the logarithmic expression gives a large and negative effect on observations that are very small (smaller than zero).

Since least-square regressions are based on assumptions of normally distributed residuals, and because our addition of 0.0001 is rather arbitrary, the results in Table 7.3 may be based on a mis-specified model. In order to test for a different specification, we experimented with a probit regression model. This model estimates the (marginal) effects on the *probability* (not the propensity) that two regions cite each other by a (marginal) increase in the explanatory

impact of distance decreases over time.[20] This study does not allow any conclusions on this question.

Third, the data reveal that knowledge flows more freely within than across national borders. Intra-country spillovers are more extensive than inter-country spillovers. Thus the concept of national systems of innovation seems to be relevant for technological competencies among European regions. However, the impact of national systems of innovation may be reduced by developments in communication technology and by European integration. Thus future research should investigate whether the impact of country borders has decreased over time.

Fourth, productivity gaps play an important role in the spillover process through their impact on absorptive capacity. Spillovers are most effective when the receiving region lags somewhat (but not too much) behind the spillover-generating region. Also, we find that regions that patent in approximately equal amounts (that is, regions that are on approximately equal technology levels) share most spillovers. This result is to be taken as a confirmation of the result from technology gap theories that technology diffusion is in no sense automatic, but demands a certain level of economic development, in addition to innovative efforts and favourable institutional settings. In particular, this shows that in the European context, spillovers are mostly taking place between a limited set of already fairly highly developed regions.

This leads us to the conclusion that the European system of innovation, as far as the role of knowledge spillovers is concerned, is to be characterized as one with polarization between several centres, rather than a single system without major barriers for knowledge flows. Within these individual centres knowledge flows relatively freely, helped by relatively small productivity gaps, small geographical distances, absence of national borders and similar or complementary specialization patterns. Across these centres, and between these and more peripheral regions, there are much fewer technology spillovers.

APPENDIX 7.1

Patents

Data on patents are from the European Patent Office (EPO, 1996). The address of the innovator(s) is used to assign regions (NUTS level 2) to patents. In the data provided by EPO each patent is assigned to one main technology class. Based on the concordance table in Verspagen, van Moergastel and Slabbers (1994) these technology classes are assigned to 22 economic sectors. Thus, the EPO patent data are used to construct data on patenting in 112 European regions, including Ireland, Norway, Denmark and Finland for which regional breakdowns are not available. In the data used in this chapter one mid-Swedish region is lacking. EPO also provides data on patent citations. To preclude or limit patent protection, the patent offices search previous patent applications in the same and related technology classes, and refer to relevant existing patents. These references constitute links between related patents and are thus used as indicators of technological spillovers..

Geography

There is no coherent data available on distance between European regions. For the purpose of this chapter, such data were constructed on the basis of maps of European regions from Eurostat (NUTS level 2) (Eurostat, 1995). The distance between two regions was set as the smallest number of regions one has to cross to reach one region from another one. Thus intra-regional distances were set equal to zero, the distance between two adjacent regions was set equal to one and so on. In the case of sea separating two regions a 'dummy region' was constructed. Thus the distance from, for instance, French regions next to the English Channel and the corresponding English regions was set equal to two. The procedure yielded a list of distances between all pairs of the 113 European regions (one Swedish region is lacking in the other data used).

Economic Development

The data on economic development are regional (NUTS level 2) GDP in PPP per habitant in ECU from 1992 and are taken from Eurostat (1997). GDP data from Norway are from Statistics Norway (1997).

Regional Compatibility

The index for regional sectoral compatibility between two regions (regions i and j), s_{ij}, was calculated in the following way. The starting point is a sector-by-sector matrix Z which describes the sectoral citation relations. In this matrix, the element Z_{pq} denotes the number of patents originating from sector p cited by sector q. We construct a new matrix z by dividing the elements of Z by the column sums, that is, $z_{pq} = Z_{pq}/\Sigma_p Z_{pq}$. The matrix z describes the distribution of a sector's received spillovers over spillover-generating sectors. For each region i, we now calculate the share of sector p in total patenting as $\sigma_{ip} = P_{ip}/\Sigma_p P_{ip}$, where P is the number of patents. The next step is to calculate, for each region, 22 (that is, the number of sectors) correlation coefficients ρ_{ip} between z_{pq} and σ_{ip}. Now calculate the share of a region in patenting of sector p as $\chi_{ip} = P_{ip}/\Sigma_i P_{ip}$. The regional sectoral compatibility between regions i and j is now calculated as the correlation coefficient between the 22 observations on ρ_{ip} and χ_{jp}. This correlation coefficient measures to what extent the sectoral patenting structure of region j is likely to be cited by region i, given the sectoral structure of i and the sectoral citation linkages. The range of the compatibility index is between minus one and one, where minus one denotes that there is no probability of region i citing region j, and one means that the sectoral distribution in patenting in the two regions is perfectly conducive to citation. The matrix of regional compatibility indexes was stacked into a list for all pairs of regions. Note that this measure of regional sectoral compatibility is not symmetric so that, generally, $s_{ij} \neq s_{ji}$.

NOTES

* We thank Marjolein Caniëls for help in constructing the distance data used in this chapter. We also thank Jan Fagerberg, Bart Los, Eddy Szirmai, Erik Biorn, Karl Ove Moene, Arne Melchior and participants at the TSER workshop in Gothenburg, 4–6 June 1998, and at a seminar at the University of Linz (Economics Department), June 1998, for useful comments.

1. Grossman and Helpman (1991, 1995).
2. Marshall (1948, Chapter X).
3. The other two were local markets for specialized skill (labour market pooling) and for specialized intermediates.
4. Vernon (1966, p. 192) states that: 'There is good reason to believe, however, that the entrepreneur's consciousness of and responsiveness to opportunity are a function of ease of communication; and further, that ease of communication is a function of geographical proximity.'
5. Krugman (1979) constructs a model of technology gaps in which laggard countries continuously take over old-fashioned products developed in the most advanced countries, which gives rise to a product cycle theory in the Vernon fashion. Krugman (1986) extends this technology gap theory and demonstrates that catch-up may harm the most advanced countries, while technological progress on the frontier increases income in both advanced and developing countries.
6. It is beyond the scope of this chapter to examine the (dis)advantages of patents or R&D indicators in detail (see the survey by Griliches, 1990).
7. In fact, it might be the case that the inventor lives in one region, but works in a different region. (This may be particularly so when inventors are well-paid employees who can afford to live in nice locations at a relatively far distance from their workplace.) However, given that the regional grouping we use consists of rather large geographical areas (often NUTS-I level), this problem is unlikely to be severe. To assess its impact, we also calculated the numbers of patents based on applicants rather than inventors, and the correlation between the two measures was high.
8. Note that Greece appears more north on the map than it actually is. This is done for typographical reasons.
9. High-tech sectors are usually defined on the basis of R&D intensity (see OECD/EUROSTAT, 1995). The sectors pharmaceuticals (3522), computers and office machines (3825), electronics (3832), aerospace equipment (3845) and instruments (385) are usually considered as high tech (ISIC rev 2 numbers between brackets). We adopt this definition in Table 7.1.
10. Formally, the Herfindahl index is defined as $\Sigma_i x_i^2$, where i indicates regions and x_i is defined as $X_i/\Sigma_i X_i$ (X denotes the number of patents).
11. Studies estimating the impact of so-called indirect R&D embodied in traded inputs on productivity (e.g. Coe and Helpman, 1995; for a survey see Mohnen, 1992) are generally within this interpretation of spillovers.
12. Data for R&D-financing links are available at the level of institutional sectors within countries. Thus, for example, one has information on which part of business R&D is financed by government in a particular country. At the regional level, however, these data are not available. Moreover, even the data on the institutional shares in regional R&D are so incomplete as to prevent us from using them in a sample as wide as the one we have in Figures 7.1 and 7.2.
13. In principle, this list can be blown up to take into account economic sectors, that is, a region-by-region by sector-by-sector list. Obviously, this would increase the number of observations dramatically (more than seven million), so only selective use will be made of the sector dimensions, and most of the work will concentrate on the regional dimension. Verspagen (1997a, 1997b) analyses sector-by-sector citations.
14. The correlations between (the log of) the number of citations between two regions and (the log of) these regions' individual number of patents are very close to 0.50.
15. We also experimented with an alternative index that reflected overlap in industrial technological specialization between regions (the exact index was defined as the sum of the absolute

value of differences between sectoral shares in patenting). As the compatibility index captures technological linkages also between different industries, this index was preferred. The regression results were not sensitive to the index used.

16. All reported estimations exclude intra-regional spillovers. These are excluded because we have no way to distinguish between intra-regional, extra-firm spillovers, and intra-regional, intra-firm spillovers, which makes the interpretation of such regressions difficult. We have experimented, however, with regressions including intra-regional spillovers, and this did not change the results markedly. These results are available from the authors on request.
17. Suppressing the other variables, spillovers depend on distance according to the following expression: Spillovers = $Adij^{-0.37}$.
18. This result indicates that country borders significantly affect the *intercept* of the regression line. We tested whether this applied also to the *slope* coefficients of the other variables, but it was not possible to demonstrate any clear effects. Detailed results are available from the authors on request.
19. We tested whether the value of the ln*GAP* variable for which the top of the parabola occurs is significantly different from zero by using a non-linear Wald Test. The null hypothesis (that is, that the maximum value occurs at ln*GAP* = 0) is rejected at the 10 per cent level, but not at the 5 per cent level ($P = 0.1$, $F = 2.71$).
20. See also Verspagen and De Loo (1998) for a discussion of the time dimension in European patent citation statistics.

REFERENCES

Abramovitz, M. (1985), *Catching Up and Falling Behind*, Economic Research Report No. 1, Stockholm: Trade Union Institute for Economic Research.

Caniëls, M. (1996), *Regional Differences in Technology. Theory and Empirics*, MERIT Research Memorandum, 2/96–009, Maastricht.

Coe, D.T. and E. Helpman (1995), 'International R&D Spillovers', *European Economic Review*, **39**, 859–87.

European Patent Office (1996), *European Patent Bulletin Information on CD-ROM*, Wien.

Eurostat (1995), *Regions – Nomenclature of Territorial Unit for Statistics*, Luxembourg.

Eurostat (1997), *Regional Statistics*, Electronic Version, Luxembourg: New Cronos Systems.

Fagerberg, J. (1988), 'Why Growth Rates Differ', in Dosi, G. et al. (eds), *Technical Change and Economic Theory*, London: Pinter, 87–99.

Fagerberg, J. (1994), 'Technology and International Differences in Growth Rates', *Journal of Economic Literature*, **XXXII**, 1147–75.

Fagerberg, J. and B. Verspagen (1996), 'Heading for Divergence? Regional Growth in Europe Reconsidered', *Journal of Common Market Studies*, **34**, 431–48.

Fagerberg, J., B. Verspagen and M. Caniëls (1997), 'Technology, Growth and Unemployment across European Regions', *Regional Studies*, **31** (5), 457–66.

Griliches, Z. (1979), 'Issues in Assessing the Contribution of Research and Development to Productivity Growth', *Bell Journal of Economics*, **10** (1), 92–116.

Griliches, Z. (1990), 'Patent Statistics as Economic Indicators: A Survey', *Journal of Economic Literature*, **28** (4), 1661–1797.

Griliches, Z. (1992), 'The Search for R&D Spillovers', *Scandinavian Journal of Economics*, **94**, 29–48.

Grossman, G.M. and E. Helpman (1991), *Innovation and Growth in the Global Economy*, Cambridge, MA: MIT Press.

Grossman, G.M. and E. Helpman (1995), 'Technology and Trade', in G.M. Grossman and K. Rogoff (1995), *Handbook of International Economics*, Vol. 3, Amsterdam: North Holland, pp. 1279–1337.

Jaffe, A.B. and M. Trajtenberg (1996), *Flows of Knowledge from Universities and Federal Labs: Modelling the Flow of Patent Citations over Time and across Institutional and Geographical Boundaries*, NBER Working Paper 5712, Cambridge, MA.

Jaffe, A.B., M. Trajtenberg and R. Henderson (1993), 'Geographic Localization of Knowledge Spillovers as Evidenced by Patent Citations', *Quarterly Journal of Economics*, **108**, 577–98.

Kaldor, N. (1978[1970]), 'The Case for Regional Policies', in *Further Essays on Economic Theory*, New York: Holmes & Meier Publishers, pp. 139–55, originally published in the *Scottish Journal of Political Economy*.

Krugman, P. (1979), 'A Model of Innovation, Technology Transfer, and the World Distribution of Income', *Journal of Political Economy*, **87** (2), 253–66.

Krugman, P. (1986), 'A "Technology Gap" Model of International Trade', in K. Jungenfelt and D. Hauge (eds), *Structural Adjustment in Advanced Economies*, New York: Macmillan.

Lundvall, B.-Å. (1992), *National Systems of Innovation – Towards a Theory of Innovation and Interactive Learning*, London and New York: Pinter.

Marshall, A. (1948), *Principles of Economics*, 6th print, 8th edn, New York: Macmillan.

Mohnen, P. (1992), *The Relationship Between R&D and Productivity Growth in Canada and Other Major Industrialized Countries*, Ottawa: Economic Council of Canada.

Morgan, K. (1997), 'The Learning Region: Institutions, Innovation and Regional Renewal', *Regional Studies*, **31**, 491–504.

OECD/EUROSTAT (1995), *Classification of High-Technology Products and Industries*, Paris: OECD.

Paci, R. (1997), 'More Similar and Less Equal: Economic Growth in the European Regions', *Weltwirtschaftliches Archiv*, **133** (4), 609–34.

Paci, R. and S. Usai (1997), *Technological Enclaves and Industrial Districts. An Analysis of the Regional Distributions of Innovative Activity in Europe*, CRENOS Working Paper, Cagliari: University of Cagliari.

Romer, P. (1990), 'Endogenous Technological Change', *Journal of Political Economy*, **98**, 71–102.

Saxenian, A. (1994), *Regional Advantage. Culture and Competition in Silicon Valley and Route 128*, Cambridge, MA and London: Harvard University Press.

Statistics Norway (1996), *Statistisk Årbok*, Official Statistics of Norway, Oslo–Kongsvinger.

Sjöholm, F. (1996), 'International Transfer of Knowledge: The Role of International Trade and Geographic Proximity', *Weltwirtschaftliches Archiv*, **132**, 97–115.

Sjöholm, F. (1997), 'Knowledge Inflow to Sweden. Does Geography and International Trade Matter?', in Fagerberg et al. (eds) (1997), *Technology and International Trade*, Cheltenham: Edward Elgar, pp. 127–39.

Vernon, R. (1966), 'International Investment and International Trade in the Product Cycle', *Quarterly Journal of Economics*, **LXXX**, 190–207.

Verspagen, B. (1991), 'A New Empirical Approach to Catching Up and Falling Behind', *Structural Change and Economic Dynamics*, **12**, 374–97.

Verspagen, B. (1997a), 'Measuring Intersectoral Technology Spillovers: Estimates from the European and US Patent Office Databases', *Economic Systems Research*, **9** (1), 47–65.

Verspagen, B. (1997b), 'Estimating International Technology Spillovers Using Technology Flow Matrices', *Weltwirtschafliches Archiv*, **133** (2), 226–48.

Verspagen, B. and I. De Loo (1998), *Technology Spillovers between Sectors and over Time*, MERIT Research Memorandum, 2/98–007, Maastricht.

Verspagen B., T. van Moergastel and M. Slabbers (1994), *MERIT Concordance Table: IPC – ISIC (rev. 2)*, MERIT Research Memorandum, 94–004, Maastricht.

Young, A. (1928), 'Increasing Returns and Economic Progress', *The Economic Journal*, **XXXVIII** (152), 527–42.

8. Changing the regional system of innovation

Bent Dalum, Magnus Holmén, Staffan Jacobsson, Mette Praest, Annika Rickne and Gert Villumsen

1 INTRODUCTION

In the broader literature on innovation, it is becoming more usual to analyse the ability to innovate and diffuse new technology using various innovation system approaches, all of which have a strong spatial dimension. In some cases the unit of analysis is a nation, in others a region within a nation. This focus on regions is not only prevalent within the discipline of economic geography but is a broader phenomenon embracing the disciplines of the economics of innovation, business strategy literature as well as the more sociological literature on industrial districts and flexible specialization (Piore and Sabel, 1984).

Achieving regional economic prosperity involves both a renewal of the dominant industries, frequently belonging to low- or medium-growth sectors, and the building of new industries, in particular in sectors characterized by high growth. Whilst the former process is difficult, it is certainly not unusual that it is successful in the sense that 'old' industries continue to exist for long time periods after they have reached their peak in terms of employment (for instance, the fishing and cement industry in North Jutland and the automobile industry and textile industries in Western Sweden).

The latter process is, however, a much more difficult one in a context where knowledge and technology development follow trajectories which are strongly shaped by history, that is, where they are path dependent (David, 1988). Not only scientific and technological competence, but also institutions and networks need to be altered or built in order to form new industries. This is a difficult process requiring a great deal of time, effort, investment and risk taking. It is nevertheless central to the economic prosperity of regions to have an appropriate balance between revitalizing existing industries and forming new ones. The purpose of this chapter is to explore the mechanisms behind the emergence of new industrial activities in high-growth sectors in two regions characterized by a dominance of low-growth industries. In particular, we will analyse the regional character of these mechanisms and, thereby, the scope for regional policy.

The chapter is structured as follows. In section 2 we will very briefly present our analytical framework. Section 3 contains a description of the current industrial profiles of North Jutland and Western Sweden. Section 4 presents four case-studies of the formation of new industries. In the final section we will go some way towards a synthesis and discuss what we have learnt about regional policy.

2 ANALYTICAL FRAMEWORK AND METHODOLOGICAL CONSIDERATIONS

In the modern literature on the economics of innovation and management of technology,[1] firms are seen as being tied together by various types of Marshallian externalities, in particular by knowledge spillovers. These may take place within user–supplier relations, between horizontally related firms through, for instance, their joint participation in a bridging institution (Carlsson and Jacobsson, 1991) or between competitors, for instance by imitation or by transfer of personnel. Firms are also seen as linked together in an indirect fashion through their dependence on a joint source of knowledge, such as universities or medical clinics.

These spillovers, and other externalities, are central to the argument for including a *spatial* dimension when the process of innovation is studied. Regionally clustered actors are argued to benefit disproportionately from spillovers due to the tacit nature of knowledge and the local nature of labour markets as well as the evolution of specific supporting local institutions and networks. This is consistent with a large amount of empirical evidence which shows that firms and innovations in the same or related industries often cluster spatially and that, consequently, regions specialize in certain technological and industrial fields (see, for instance, Audretsch and Feldman, 1994; Enright, 1995).

Less attention has been given to the process whereby the specialization within a region changes by diversification into new industries. In this process, new clusters are formed, which both presupposes and gives rise to institutional change, formation of new networks and generation of external economies.

Analysing the formation of new clusters requires an analytical framework of a system nature, which is designed for that purpose. Of those which have emerged recently, the technological system approach is particularly suitable since it advocates that the system features are, to a degree, technology specific.

This means that there are many technological systems in a country or region (for instance, one for genetic engineering and another for digital signal processing). Each technological system is unique and the systems vary in their ability to develop and diffuse new technology. Moreover, the process of cluster formation would be expected to vary between clusters formed around different technologies.

A formal definition of a technological system is (Carlsson and Stankiewicz, 1991, p. 111):

> ... network(s) of agents interacting in an economic/industrial area under a particular institutional infrastructure to generate, diffuse, and utilize technology. Technological systems are defined in terms of knowledge or competence flows rather than flows of ordinary goods and services. They consist of dynamic knowledge and competence networks.

A technological system is constituted by a number of elements:

- The level and character of *competence*, technical as well as other types of competencies, which may reside in various actors. A particularly important set of actors are 'prime movers'. These are firms or other organizations which are technically, financially and/or politically so powerful that they can initiate or strongly contribute to the development and diffusion of a new technology.
- The character of *networks*, which can be of various types. In particular, we may note networks which are conducive to the identification of new problems and the development of new technical solutions (often user–supplier networks) and more general information diffusion networks.
- The types of *institutions*, which can be 'hard' ones, such as legislation, the capital market or the educational system, or softer ones, such as culture. The role of the various institutions varies, of course. The function of some is to act as bridges between firms, between firms and academia, and in general to enable a high connectivity in the system; whereas other institutions, such as tax legislation, influence the incentive structure.[2]

In our empirical work, we apply a 'technological system approach' at the regional level, that is, we work with the emergence of regional innovation systems, which are technology specific. The approach is applied to four cases, two in each region. The cases are not identical in the two regions but are very closely related. In North Jutland, the cases involve the wireless communications cluster (mobile phones and maritime communications) as well as the medical electronics cluster. In Western Sweden, the case-studies focus on antenna and microwave technology (which is closely connected to mobile telephony) and osseointegration, which is a medical technology.

Hence, we focus on areas which are largely placed within the telecommunications and medical equipment industries which both are high-growth industries. Indeed, these were among the top six industries ranked according to growth rates in exports for the OECD countries in the period 1973–94 (Dalum, Laursen and Villumsen, 1997). In each of the cases, we will trace the evolution of new clusters focusing on the development of competence, institutional change and the formation of networks.[3]

3 INDUSTRIAL SPECIALIZATION IN NORTH JUTLAND AND WESTERN SWEDEN

The two regions studied are small. Western Sweden has well above one million inhabitants and North Jutland about half a million. Western Sweden and North Jutland have a GDP (adjusted for purchasing power parity) approximately on par with the EU average. The delineation of the regions has been done by Eurostat's NUTS 2 level for Western Sweden and NUTS 3 for North Jutland. The criterion has been the existence of a regional government authority which is somehow able to pursue an industrial policy. The two regions differ in terms of the degree of industrialization. Whereas Western Sweden is heavily industrialized with nearly twice as many employees (per capita) in manufacturing compared to the OECD average, the relative share of manufacturing employment is only 70 per cent of the OECD average in North Jutland.

As is the case in many other EU regions, Western Sweden and North Jutland are dominated by mature industries. Traditional industries[4] and metal and machinery are the largest in absolute terms. In North Jutland, these both account for 40 per cent of the total employment in manufacturing; in Western Sweden the shares are 26 and 48 per cent respectively.

In comparison with the OECD, relative employment (see RE in Table 8.1) in *North Jutland* is only higher than the OECD average in traditional industries and metal and machinery. The dominating sub-industries within the traditional industries are food processing and non-metallic industrial products (cement in the North Jutland case). Within metal and machinery, shipyards are of major importance. The latter have, though, experienced a significant reduction in employment since 1992. On the other side of the coin, North Jutland has clearly not been specialized in high-tech-oriented industries such as chemicals, electronics, pharmaceuticals and instruments.

A closer look at the sub-industries (not shown in the table) reveals some interesting growth areas. In other telecommunications (ISIC 38329) North

Jutland in 1992 had a relative employment share of more than twice that of Denmark. Information collected directly from the firms shows that employment in this segment has further increased since then. A similar increase has been seen in the printed circuits segment. Although North Jutland has usually been considered an area dominated by 'the industries of the past', the presence of a fairly broad machinery industry as well as the emergence of a dynamic electronics industry may imply that this picture is too simple.

Table 8.1 Comparison of manufacturing employment patterns in North Jutland, Western Sweden and the OECD 1991/92.

	North Jutland[a]		Western Sweden[a]		OECD	
	Share (%)	RE[b]	Share (%)	RE[b]	Share (%)	Empl. change (1975 = 1)
Traditional	40.2	1.11	26.3	0.81	36.3	0.84
Chemicals	7.3	0.47	14.0	0.89	15.7	1.18
Paper and pulp	2.0	0.68	5.5	1.96	3.0	0.99
Metals & mach. (incl. autos)	40.2	1.38	47.7	1.30	28.3	0.97[c]
(Automobiles)	1.1	0.18	20.5	3.50	6.1	0.89[c]
Electronics	8.0	0.69	5.5	0.44	11.5	1.22
Pharmaceuticals	0.4	0.40	0.1	0.30	1.0	1.46
Instruments	1.9	0.76	0.9	0.39	2.5	1.07
Total empl. in manufacturing	31 327		186 678			

Source: OECD-STAN, Statistics Sweden and Peter Maskell, Copenhagen Business School, based on unpublished material from Statistics Denmark.

Notes:
a Figures for North Jutland are 1992. For Western Sweden and OECD 1991.
b RE is relative employment compared to the OECD; OECD = 1
c Base year 1980.

In *Western Sweden*, the industrial structure is heavily skewed in favour of metals and machinery, in particular the automobile industry, with both Volvo and Saab Automobiles having their headquarters and main factories in the region. The relative employment (RE) is, as expected, very high for the automobile industry (3.5) and is also high for paper and pulp as well as for metals and machinery (1.96 and 1.3 respectively). Electronics, instruments and pharmaceuticals have, on the other hand, very low REs (0.44, 0.39 and 0.30 respectively).

The mature nature of the industrial structure of Western Sweden and North Jutland is clearly seen when the specialization patterns are compared with changes in employment in the respective industries at the OECD level 1975–91 (compare columns 2, 4 and 6 in Table 8.1). Both regions are specialized in industries with declining employment at the OECD level, and the opposite is the case in industries characterized by expansion of employment in the OECD. The industrial structure of the two regions is therefore not very conducive to employment growth. The key issue then is to unravel the process by which the regions can generate industrial activities in high-growth industries and, consequently, how to change the industrial structure to a more favourable one.

4 CASE-STUDIES IN NORTH JUTLAND AND WESTERN SWEDEN

4.1 The Telecommunications Cluster in North Jutland

Radio technology can be traced back to the 1890s when Marconi demonstrated the possibility of radiocommunications. In 1901 he managed to send a radio message across the Atlantic Ocean. The technology deals with how to propagate radio signals. The spectrum range is 800 KHz to around 30 GHz and includes signals for different purposes such as broadcasting, satellite communications and mobile communications.

This case is focused on a small cluster of high-tech firms and knowledge institutions in the field of radio technology. In 1997 a club (NorCOM) was founded by 20 firms, the Institute of Electronic Systems at Aalborg University and the local science park (NOVI). This club represents the core of an industrial cluster focused on production and development of mobile communications equipment, cordless systems, modem and fax equipment for wired and wireless systems, as well as various equipment for maritime communications and navigation, such as the Global Maritime Distress and Safety System (GMDSS), Very High Frequency (VHF) and the Global Positioning System (GPS). Figure 8.1 represents a 'family tree' indicating the historical relations between the members. In 1998 employment in the private firms was about 3700 persons in 25 firms, of which the three largest, Bosch Telecom Danmark, Sonofon and EuroCom Industries/SP Radio, employed 1400, 900 and 570 persons respectively. Aalborg University had around 12 000 students in three faculties and a staff of roughly 1200. The Institute of Electronic Systems and Centre for Personal Communications (CPK) are two large groups at Aalborg University, known for their research capacity in radio communications and speech recognition. The total academic staff in electronics at Aalborg University is more than 250 people,[5] including about 80 PhD students. CPK as a subset amounts to approximately 40 researchers.

NorCOM fits well into the characteristics of a technological system. Radio-communications technology is the common knowledge base, which demarcates the cluster. The emergence of the latter can be traced back to 1947 when Simon Petersen founded SP Radio as a consumer electronics firm (see Figure 8.1 and Dalum (1995) for further details). SP Radio managed to switch from consumer electronics to professional maritime communications equipment for small ships in the mid-1960s and conquered a leading worldwide position fairly quickly. This switch was induced by local demand conditions. North Jutland was an area with a large commercial fishing fleet and a large number of yachts. Engineers from SP Radio founded Dancom in 1973, Shipmate in 1977 and Dancall (now Bosch Telecom Danmark) in 1980. The latter was founded as a maritime communications firm but it moved quickly into mobile communications.

In the 1960s, two engineering schools emerged in North Jutland. In 1974 they became part of the newly established Aalborg University. Because the academic staff at the Technical Faculty from the very beginning was about 200 persons, the university had the potential to become a fairly powerful engineering school and technical research institution. It became visible on the national and international research scene from the early to mid-1980s. In the 1990s Aalborg University has on average produced approximately 40 per cent of the Masters (5 years) in engineering in Denmark whereas in electronics 50 per cent of the Danish Masters came from Aalborg. During the 1960s and 1970s the direct research spillover effects from Aalborg University (and its predecessors) to the local radiocommunications industry did not appear be greatly important. However, the indirect transfer of knowledge through newly graduated engineers (then Bachelors, only 3–3½ years) was important.

The 1980s witnessed important changes. The research base at Aalborg University strengthened parallel to the very fast development of the mobile communications industry. The foundation of Dancall coincided with the take-off of the NMT (Nordic Mobile Telephony standard). In the mid-1980s, Shipmate diversified and established Cetelco as its mobile communications subsidiary. The emergence of the pan-European and fully digitalized GSM (Global System for Mobile Communications) standard from the late 1980s required an innovative effort on a considerably larger scale. With this background the direct competitors, Dancall and Cetelco, established a pre-competitive joint venture company, DC Development, with the aim of jointly developing some of the core technologies in a first generation GSM terminal. The latter could then be put on the market under different brand names (with, for instance, various designs and functional features) by the two competing firms.

At the same time Aalborg University, the regional county council and a major local bank established a science park, NOVI. Although these two efforts were independent, they were, by coincidence, well timed. NOVI needed a high profile prestige project to become recognized and Dancall and Cetelco, still direct competitors as far as their final products were concerned, needed 'neutral ground' near the university. In 1992 DC Development, through its mother companies, managed to launch a first generation GSM terminal at the same time as a few world leaders, such as Motorola, Ericsson and Nokia.

In 1992 SP Radio had succeeded within a major development effort, a Global Maritime Distress and Safety System (GMDSS) station. The firm was among the first of two to deliver total GMDSS systems; its main competitor was Japan Radio Corporation. SP Radio retained its technologically leading position during the 1970s and 1980s and moved from the smaller ships segment to large commercial vessels.

Since 1992 major changes have occurred. Foreign take-over of a considerable part of the cluster has taken place. In *mobile communications*, the DC Development joint venture had to be closed down in 1992. The two mother companies, Dancall and Cetelco, could not manage financially with this project, which – from a technological point of view – was highly successful. A knowledge asset of 30 young electronic engineers, who succeeded in launching a major technological innovation as a team, had to be split. The creation of this critical asset for the mobile communications industry in the region had strained the financial situation of the two mother firms to such an extent that they lost their independence. Although the team of engineers was split, it managed, however, rather quickly to 'clone itself' in that several firms in the region were able to develop GSM terminals shortly after the collapse of DC Development.

In 1993, Dancall was acquired by Amstrad (UK), and in 1997 by the German Robert Bosch Group which has invested in a significant expansion of the production capacity. Bosch built a new factory in autumn 1998 with a production capacity making it possible for Bosch to be among the five biggest firms in the world. Cetelco was gradually taken over by the German telecommunications firm Hagenuk from the late 1980s and in 1998 Italian Telital took over the firm.

An important feature of the cluster has been the continuing emergence of new small development firms focused on developing equipment, often for large multinationals. A group of engineers from Dancall founded T-Com in 1987. This was taken over in 1991 by the South Korean electronics multinational Maxon (8000 employees). From an initial size of around 10 employees, Maxon Denmark has grown rapidly to 100 employees, and it now has a strategic position in product development within the Korean company. In the 1990s, two other such new 'Schumpeter Mark I' firms have emerged, both of them as spin-offs from the established larger firms; ATL in mobile and RTX in cordless communications.[6]

Apart from Nokia's recent significant expansion in Copenhagen, North Jutland has emerged as the Danish centre for the development of wireless

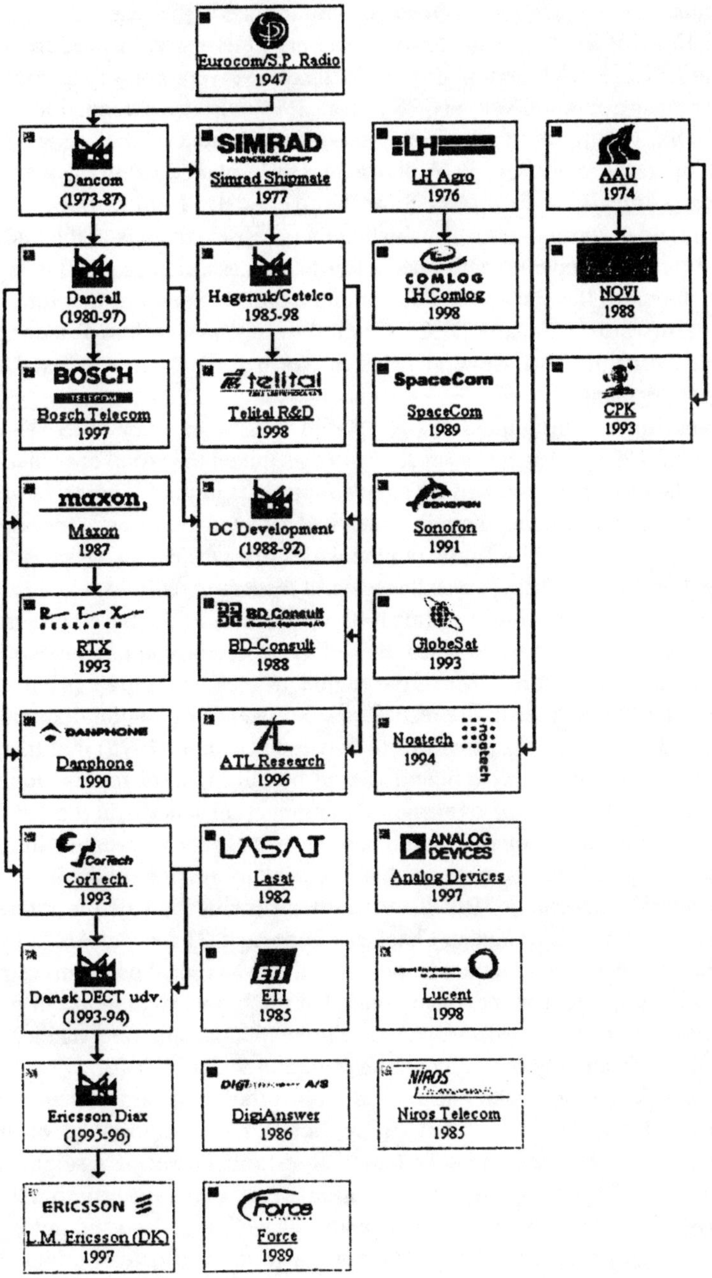

Figure 8.1 The NorCOM history

communications equipment. Of major importance is the research capacity of the Institute of Electronic Systems at Aalborg University. In 1992 the Danish Technical Research Council allocated a large five-year grant to a new Centre for Personal Communication (CPK); from 1997 this was extended for another five years. Of importance was also the establishment of the second Danish GSM operator, Sonofon, in Aalborg in 1991, owned jointly by a Danish telecommunications firm (GN Store Nord) and Bell South (US). Another related field is cordless phones, which are wireless extensions of the traditional cabled (wired) telephone networks. After its successful implementation of the GSM standard, the European Telecommunications Standard Institute, ETSI, has put forward a standard, DECT (Digital European Cordless Telecommunications), for this field. Several firms in North Jutland have focused on the cordless segment.[7]

In maritime communications SP Radio was taken over by SAIT-Radio Holland in 1992. Later the German STN Atlas joined the group of owners. This foreign take-over was caused by personal motives in the founding family and not by the kind of financial strain which had affected the firms in mobile communications. In the 1990s EuroCom Industries (the new name) has consolidated its position as one of the two to three world leaders in its field.

The history of the cluster points to the capability of Aalborg University to deliver engineers and basic research with a sufficiently application-oriented touch to serve as a core asset of the region. A technical university may thus play a rather direct role in the restructuring process of a region dominated by traditional industries. It seems fair to conclude (Dalum, 1995) that this high-tech cluster could not have been fostered by pure market forces alone. The deliberate political struggle for the first engineering schools in the 1960s and again a major battle for the establishment of Aalborg University in 1974 – against many odds at that time, such as the peak of the first oil crisis – were of fundamental importance. But also the specific character of the interaction between university and industry has been important. The principles of project-based learning, often with the solution of real-life technical problems as part of the students' project work, have created skills highly demanded in product development intensive firms. A general entrepreneurial spirit among electronics engineers was already prevalent in the region in the late 1960s.

What has distinguished the region from others appears to have been a widespread awareness, explicit or implicit, of the importance of further developing the technological system. The establishment of a science park, NOVI, in 1988 was again an indication of a successful combination of political efforts by various individual actors as well as private and public institutions. A fortunate mix of new bridging institutions (hosted by NOVI) and financial support schemes, such as various EU programmes (for instance, the Target 2 and Stride) and various support schemes from the Danish Ministry of Business and Industry, has gradually evolved into a fairly coherent policy framework. The importance of the collective consciousness regarding the existence of a de facto technology system should not be underestimated as an unwritten framework or as a set of intuitive guidelines for many single-policy schemes supporting the system, such as tailor-made training programmes of shop floor workers.

4.2 An Emerging Electromedical Cluster in North Jutland?

The central technology in the North Jutland activities in this field is functional electrical stimulation (FES), which is based on use of the intact functionality of the muscles after paralysis (for instance, caused by strokes or accidents). Based on the neural coding in the human sensory-motor system together with new technologies for registration and processing of electrical signals from muscles and nerve tissue, it is possible to activate the paralysed muscles. Related technologies are motor control, dealing with the interaction between motor and sensory control mechanisms in the central nervous system, and pain research (control).

The field is only an embryo of a potential industrial cluster. Taking our point of departure in the Jacobsson and Rickne (1996) approach for analysing the relation between a technological system and new technology-based firms, some of the elements for a regional electromedical technology system exist, others do not. The component which is necessary although it is not sufficient in itself – a research base of a certain critical mass – has emerged during the last 10–15 years at Aalborg University. The Medical Informatics Section at the Department of Medical Informatics and Image Analysis (MIBA), itself a part of the large Institute of Electronic Systems, has a research staff about 25 (PhD students included).

MIBA has fostered a fairly strong research unit, SMI (Centre for Sensory-Motor Interaction) focused on biomedical engineering research in neuroscience – an outcome of activities initiated in the late 1970s. SMI has a staff of roughly 30 people and hosts the International Doctorate School in Biomedical Science and Engineering with about 20 PhD students, funded by the Danish National Research Foundation. The research has resulted in electrical stimulation concepts which, on an experimental basis, have been transformed into handicap aid devices tested on people damaged by, for instance, strokes or traffic accidents. Applications for foot control for people with a drop-foot indicate that natural sensory signals can be used to control paralysed muscles. This kind of technology has formed the basis for a small start-up venture, Neurodan, in which NOVI as a venture capitalist has invested seed money, and the venture has also received funds from the EU Target 2 programme. Neurodan has applied for patent protection of important parts of its technologies.

The decisive next stage is then what kinds of companies are prepared, as licensees, to transform innovations from small research-based ventures, such as Neurodan, to commercially marketable products.[8] At present the number of university spin-offs in the electromedical field is very small, but the future potential appears to be fairly large. The research base, the supporting and bridging institutions and a seed capital institution are available. The missing link – and the core problem – is lack of incumbent firms. There are only three electromedical firms in the region: Artema Mec, a Swedish-owned producer of monitoring and emergency care equipment (100 employees), Cortex Technology (12 employees) in ultrasound scanning equipment for dermatologic diagnosis and Judex (15 employees), a university spin-off concentrating on equipment for analysis of signals from nerves and muscles.

Denmark is well known for a series of small- and medium-sized world leading niche firms in electromedical equipment, but most of them are located in the Copenhagen region. The geographic concentration of public research as well as private firms in the fields of medicine, pharmaceuticals and biotechnology around Copenhagen is rather pronounced. Apart from the lack of incumbent firms in North Jutland, there is no medical faculty and no university hospital. But there is increasing cooperation between the university hospital in Aarhus and the local hospital in Aalborg, moving towards giving Aalborg formal status as an affiliate to Aarhus.

The potential of North Jutland is based on strong university research capacity in some specific segments well suited for interdisciplinary collaboration between hospitals and electronic engineers. The SMI research team has been able to compensate for the lack of medical institutions in the region through intense collaboration with the other university hospitals and medical faculties in Denmark. Although there is some room for manoeuvre here as far as national industrial policy is concerned, the core problem in fostering an electromedical cluster in North Jutland is the lack of incumbent firms. While the NorCOM cluster originally was triggered by private firm initiatives, an electromedical cluster will have to be the outcome of a public R&D push-oriented process.[9]

4.3 The Technological System of Antenna and Microwave Technology in Western Sweden

Antenna technology deals with how to send and receive electromagnetic waves (EM). Antennas are only used in products and systems together with complementary technologies. Some of these, for instance high-frequency amplifiers or wave guides, are related to the antennas by their close spatial proximity in product design and the large technical knowledge overlap. These technologies are referred to as microwave technologies. The system of antenna and microwave technologies in Western Sweden is delineated to include actors which are active in the generation and the diffusion of these technologies.[10] They are used in a diverse range of products such as radar, microwave ovens and mobile phones.

Since the Second World War Western Sweden has had a strong position in antenna and microwave technologies, but the number of substantial actors has never been more than four or five. The economic importance of the technology was already substantial in the 1960s, and increased significantly in the 1990s as a consequence of the rapid expansion of mobile communications. There are more than 5000 employees in the private sector in the field (compared with 1200 in the mid-1960s and 2000 in the early 1980s), which contributes significantly to the renewal of the industrial base of the region.

The formation of this technological system is driven by three factors. The first is the character of the specialization at the Department of Electrical Engineering at Chalmers University of Technology. The formation of the Onsala Space Observatory (OSO) led to rather specific demands on the research at the university, such as the development of extremely sensitive receivers or tailor-made antennas. The result was a strength in a broad range of antenna and microwave technologies. OSO had, and still has, the role of a competent and demanding customer. This research profile has clearly been transplanted to the undergraduate teaching at the university. A substantial number of people over the decades have graduated from Chalmers with a relatively deep and broad knowledge in antenna and microwave technology. The concentration of this broad range of skills in the region is unique in a Swedish and, presumably, even in an international context.

The second factor was the establishment in 1956 of an LM Ericsson affiliate in the region, now known as Ericsson Microwave Systems (EMW). The basic motive was to get access to the local pool of engineers, in particular the new graduates from Chalmers. Because of its size, the presence of Ericsson dominated the regional activities of the field from the very beginning. Originally, EMW was concentrating on military applications, especially radar. However, diversification into non-military applications (for instance weather radar or microwave links) was also initiated. Although many of these have been technical successes, no major economic breakthroughs were seen until the rapid expansion of mobile communications in the 1990s.

The third factor in the formation of the technological system was the merger in the beginning of the 1990s of the satellite applications activities at Ericsson with Saab Space, which made, for instance, on-board computers for satellites. The four actors, Chalmers University of Technology, Onsala Space Observatory, Ericsson Microwave Systems and Saab Ericsson Space (the name of the merged activities) form the regional core of interdependent actors with significant knowledge overlaps and areas of common interest.

The number of core actors in the region was increased further in 1997 by the establishment of Allgon, which has a substantial share of the world market for antennas for mobile communications. Its investment in the region (like Ericsson's in the 1950s) was made in order to gain access to the regional labour pool, especially key personnel from Ericsson and the graduates from Chalmers.[11]

Apart from the regional *core* of actors, there have been a number of other actors, such as individual inventors, spin-offs from Ericsson and Chalmers and other firms. The most important of these is Saab Marine Electronics, which produces level measurement systems for, for instance, supertankers. The firm does not belong to the core since its skills in antenna and microwave technology are more limited, at least according to the core actors who have not found it necessary to develop close links with the firm. Partly for this reason, Saab Marine Electronics has sourced antenna and microwave technologies from outside the region.[12]

The key issue today is how the region should maintain and exploit its competitive advantage in antenna and microwave technology, an advantage based on its early entry into the field. In spite of the significant accumulation of human capital, all local actors face a considerable obstacle to expansion – lack of skilled personnel. This means that the region cannot fully exploit the market opportunities in mobile communications, and, thus, further speed up the process of regional renewal. An experienced engineer in one of the core firms claimed that 'if there was a region in the world where 30–40 microwave engineers graduated every year, we would be established in that region'. Indeed, it was recently announced that Ericsson Microwave Systems has started an R&D unit in Milan in order to exploit the competence base there. Institutional change leading to a substantial increase in the number of degrees awarded at the undergraduate and graduate level is required.[13]

There are currently several changes going on at Chalmers, which address these problems: (i) the number of undergraduate students taking courses in microwave technology has increased from around ten to 40 in the last five years but a ceiling on size has now been placed by the university; (ii) since supervisory capability is a scarce resource, Ericsson Microwave Systems is trying to address this problem by having experienced engineers supervise and mentor not only undergraduate students but also PhD students; (iii) 'competence centres' have been set up. These are R&D centres, funded by the Swedish National Board for Industrial and Technical Development (NUTEK), with the aim of building networks of firms and academics. One example is the Centre for High Speed Technology at Chalmers (CHACH), a research centre for high-speed electronics; (iv) a perceived emerging technological discontinuity into 'intelligent antennas'[14] has led Ericsson to make a major investment in research at Chalmers.[15]

Hence, whereas the region seems to be well equipped in terms of its knowledge base to further expand its activities in antenna and microwave technology, it is worrying that the volume of activities (undergraduate education and supervisory capacity) still remains relatively low. The region has a potential for greater expansion in the field, but some evident opportunities will be left for exploitation in other regions (probably in other countries).

4.4 The Technological System for Osseointegration in Western Sweden

In the 1950s the Western Swedish Professor Brånemark discovered that titanium is accepted by the human body, and can be integrated with living bone and tissue. Titanium could therefore be used as a biomaterial, that is, a material used to treat, enhance or replace human functions. The field is labelled osseointegration – the way in which bone healing takes place so as to integrate metal with living bone.

This discovery opened up opportunities for research on a variety of applications. The key was the development of the titanium screw, serving as a fixture into the human bone for a range of prostheses.

Four different sets of applications have been exploited in the region; dental implants, hearing aid implants, facial implants and orthopaedic implants. The Brånemark team started trials on *dental implants* in 1965. In the late 1970s, all major development was completed and the team was ready for a broader industrial cooperation. The Institute for Applied Biotechnology was established in 1978, with the goal of becoming a broad competence centre concentrated on osseointegration, and to bring products to the customer. In the same year, the technology was licensed to a large Swedish firm, Nobel Industries.[16] This licence included exclusive rights to the technology for all applications above the clavicle, while the Institute for Applied Biotechnology retained the rights to use the technology for all other parts of the body.

Nobel Industries had no previous experience of implants but wanted to diversify into the medical field through establishment of a new venture, Nobel Biocare.[17] Located in Gothenburg, and with production facilities near the parent company in a different region, Nobel Biocare benefited greatly from research at the local University Hospital Sahlgrenska, the Institute for Applied Biotechnology and Chalmers University of Technology. Nobel Biocare went from a financial loss in 1985 to being a profitable company with over 700 employees and sales around 133 million USD in 1997, and with the recent acquisition of one of the leading American implant companies, Steri-Oss, these figures have been considerably increased.[18]

In the region, a second diversifying firm, Astra Tech, a subsidiary of the large pharmaceutical company Astra, entered into dental implants in the mid-1980s, partly by way of recruiting an R&D director from Nobel Biocare.[19] The competence from Nobel Biocare thus spilled over to the new player who, so far at least, is a more marginal actor.

In the mid 1970s Brånemark and his team initiated cooperation with the Department for Applied Electronics at Chalmers University of Technology, to develop a bone-anchored hearing aid. In 1984, the two central researchers at Chalmers formed a university spin-off, P&B Research. As Nobel Biocare had the patent for the Brånemark method, this company was perceived to be the best outlet for the application and a licensing agreement was set up where P&B Research should carry out all development. In 1997 the product area gave Nobel Biocare 28 million SEK in turnover.[20] Today, the licensing agreements have expired and the two firms, Nobel Biocare and P&B Research, act in parallel on the market, while P&B Research, which is still a very small company, is developing the next generation hearing aid.

Osseointegration can be used for replacing orthopaedic functions; knees, legs, fingers, etc. that may be in need of replacement. The Institute for Applied Biotechnology has continuously worked on a number of different applications, building up invaluable experience. This competence base has been created in cooperation with other researchers, in Gothenburg, Lund, and also in La Hoya, USA, too. The orthopaedic field is now ready for exploitation, after 15 years of competence building. The market potential is large, probably at least as large as for the dental application.

Research on finger joints, where the titanium screws are inserted and attached to a silicon spacer, has also been carried out at Sahlgrenska since the early 1980s. This research team, drawing on the Nobel Biocare experience, established a formal contract with Astra Tech in the early 1990s. The project is now well on its way, and Astra Tech seems to be sufficiently patient to wait for the product. All of these firms will probably enter in an early phase of the market, and thus the region may well come up with yet another set of early movers.

The first attempts at using titanium implants for facial prostheses were made at Sahlgrenska in the late 1970s. Since then ears, eyes and facial prostheses have been given to thousands of patients. However, even though the licences to use the technology for this application belong to Nobel Biocare, the company has not focused on this niche, due to a smaller market potential. These endeavours are therefore still on a smaller scale and dependent on the initiative of the university hospital and the Institute for Applied Biotechnology.

The region has performed fairly well. In three of the applications, there are first movers from Western Sweden. In dental implants Nobel Biocare has already managed to acquire a good market share, and there is a large potential in the other applications. It is the specific academic competence which has given the region much of its competitive edge in this science-based industry. Networks between the various university departments and between these and industry, as well as the emergence of a regional labour market, have contributed further to localizing much of the activity in the region. The issue at stake is whether the region can continue to build on this strength. Three observations can be made in this context.

First, national science policy has supported research in osseointegration for many years.[21] The field of biomaterials (of which osseointegration is a part) is judged as strategic and a consortium involving researchers from a range of academic disciplines and firms within many applications has been set up. In this way, the consortium has provided a national (and regional) platform on which new networks can be built and old ones strengthened. Hence, an institutional change strengthening the competence base has taken place.

Second, the attitude of the universities to commercialization is important. In the dental implant case, Brånemark pushed the industrialization *in spite of* the attitude of the university. Today, the universities in the region have all taken significant steps towards a more helpful attitude, the most recent example being an innovation park set up at the Sahlgrenska University Hospital. Again, institutional change favours a further exploitation of the knowledge field in the region.

Third, it is likely that the region will benefit from having more than one firm involved in each application and more firms exploiting new applications, such as facial and orthopaedic implants. In addition to further diversification by Nobel Biocare and Astra Tech, the new academic spin-offs in the orthopaedic field would seem important to the region. Still at an early stage, these firms are able to find financing (within or outside of the region) and seem to have a viable technology. However, in the competition between regions, where Western Sweden and La Hoya, California are two of the leading centres for research regarding osseointegration, it remains to be seen where these companies will be located.

5 CONCLUSIONS AND POLICY DISCUSSION

The case-studies show that, perhaps surprisingly, we have witnessed a process of emergence of new technological systems where many of the mechanisms have a strong local dimension. The origin of the process is in all cases local in character. In the case of both osseointegration and antenna technology, it was free university research which was the driving force behind the new technology. In North Jutland, the origin of the mobile phone and maritime communications firms can be traced back to SP Radio and its efforts to find new applications among local customers in the large local fishing fleet; whereas the embryonic medical technology cluster has its origin in free university research. Hence, the origin lay in local research and experimentation.

What the regions have in common is that there are locally constrained mechanisms for enlarging the initially small competence and actor base in the emerging systems. A *first* mechanism lies in graduate teaching where the students of pioneer researchers choose to make a career in local universities creating a specialization in the regional competence base for both research and teaching. A *second* mechanism is the strong local networks between the firms and between industry and university which may act as a vehicle for knowledge transmission. A *third* is the emergence of a local labour market for specialists. A *fourth* lies in institutional change allowing for a change in the supply of scientists, engineers and PhDs, in the provision of risk capital, and the formation of meeting places. *Fifth*, spin-offs form a locally constrained mechanism for enlarging the actor base. This is most evident in the case of mobile and maritime communications in North Jutland but to a lesser extent it is also true for Western Sweden. In that region, diversification by existing firms in the case of osseointegration (Astra Tech) adds to the mechanisms for enlarging the actor base.

Whereas these mechanisms are locally constrained, national and international linkages of many kinds form an integral part of the emergence of the regional technological systems. The most obvious mechanisms lie in direct investment by actors outside the region. This was of central importance to the two cases in Western Sweden (Nobel, Ericsson) as well as in the mobile phone case in North Jutland (Bosch, Ericsson, Maxon). In all cases, however, it was the unique regional knowledge base which attracted the firms, so the 'foreign' direct investment reacted to, and greatly strengthened – rather than initiated – a process whereby new systems emerged. With the 'foreign' direct investment, the region received an injection of assets which were complementary to its knowledge base (capital, management, reputation in the market, technology, etc.). Bosch Telecom Danmark is integrated in the more basic R&D activities of Bosch (in Germany) and the previous Cetelco will be integrated in the R&D activities of Telital (in Trieste).

National and international linkages abound in the sourcing of knowledge and technology. Whereas the regional context constitutes an importance source of knowledge and competence, the firms are part of national or international technological systems, which are of vital importance to them. Nobel Biocare has, for instance, strong connections with dentists abroad, Ericsson with other Ericsson units in its multinational organization as well as with other academic centres, and, finally, the regional scientific specialization is very dependent on national science policy actors, mainly located in the respective capitals.

After this brief synthesis, let us expand on two themes of relevance to policy.

(i) The central role of universities (science and technology)

The universities have a number of functions to fulfil in the process of regional transformation. Within the universities, much experimentation is taking place, including exploring new applications of a given technology. Researchers are important in the networks and universities are the source of highly trained people, at both the undergraduate and graduate levels, as well as a source of new firms through spin-offs (nearly all spin-offs in Sweden are found around universities (see Saemundsson et al., 1997)). All these functions are found in the case-studies. There are then a number of issues which need to be raised in the context of policy.

First, the cases have shown the value of free research, that is, academic research not directed by industry. The distinction between free and targeted research is quite different from that of basic and applied research. Both free and targeted research can be basic *and* applied; osseointegration being an example of free research that was first basic and then applied. The issue at stake here is the appropriate balance between free and targeted academic research. In the Swedish case, recent changes in the context of science policy have tilted the balance to targeted research, which may mean that the opportunity for experimentation that goes beyond the visions of current industry diminishes. Although part of the research in radiocommunications at CPK/Aalborg University has the character of targeted research (directed towards specific applications), a balance between the latter and the free search for knowledge appears to have been established as de facto guidelines for the choice of research projects. In the SMI case it is evident that the development has been driven by the principles of fundamentally free university research.

Second, the attitudes of the universities towards commercialization of academic results have a bearing on the exploitation of the regional competence base. These attitudes may take many forms. Are the researchers encouraged to commercialize and rewarded for these efforts? Are significant resources devoted to technology-licensing offices? Are places provided where researchers can meet industry and venture capital firms? Does the university work to make sure that there is venture capital available? Are there courses and back-up facilities (for instance, patent officers) for the researchers, etc.? Hence a policy issue of significant standing is the importance given to commercialization by the universities. This is, of course, primarily a policy issue for the university, which is therefore a key regional policy actor.

Third, as and when there is a demand for a new technology on a large scale, the region needs to have universities which have an interest in and an ability to scale up the 'production' of engineers and scientists. This may imply that there is a need for institutional flexibility as well as diversity (Carlsson and Jacobsson, 1997). It is not enough to have universities which cater for the 'elite' of the students graduating from secondary school; 'volume' production of, say, computer scientists, is incompatible with an 'elitist' education. A regional policy issue is then the extent to which there is adequate flexibility and diversity of its higher educational system.

Fourth, in science-based technologies it is particularly important that the universities succeed in their recruitment of academic leaders. In Sweden, which often still insists on having only one professor in each subject, the wrong choice of professor can influence the whole process of specialization of the competence base, as can a good recruitment. Flexibility at the universities includes the ability to rectify initial recruitment mistakes in order not to block the development of virtuous circles. One successful professor may set in motion a process where volume research is pursued and a large number of PhD degrees are awarded with consequent opportunities to shape the process of regional renewal (as with Professor Brånemark in osseointegration). The Aalborg University-related cases do show much less vulnerability to the potentially negative effects of non-productive but still powerful individual researchers. The size of the departments and apparently a 'flatter' decision-making process appear to have been among the key factors.

In conclusion, the universities have a key role in the process of regional transformation; they must consider their responsibility as policy agents – and act accordingly.

(ii) The diverse mechanisms to exploit the knowledge base

The evolution of the four systems has been quite different, illustrating that there is not only one path to tread to attain regional renewal but several. In particular, the constellations of actors can vary from the Western Swedish dominance of large firms to the Jutland experience with clusters of smaller firms. In particular, it is interesting to contrast the Western Swedish case in antenna technology with the North Jutland mobile phone case. In the Swedish case, Ericsson is by far the most important actor and there are only a few other firms. Ericsson is extremely dynamic, attractive for new engineers and absorbs a large share of the newly graduated engineers and PhDs, which, of course, leaves very little room for smaller firms. North Jutland, on the other hand, is dominated by a cluster of smaller firms, which is an archetypal example of the Marshallian industrial district.

The different patterns have their strengths and weaknesses. The Jutland experience, with the local cluster being acquired by foreign firms, suggests that it is dangerous to rely purely on technological capabilities and, therefore, to lack complementary assets. The Western Sweden experience points to the risk of having too few actors since all actors need to focus their own research, as Ericsson and Nobel Biocare certainly do, in a limited area of application where there remain vast opportunities. Few actors may therefore mean that the competence base of the region is not fully exploited, as is also the case for medical electronics in North Jutland.

There are two mechanisms for enlarging the number of actors. First, 'foreign' direct investment may supplement regional actors if the former is not only done through acquisition of existing firms. Foreign direct investments may, as in the Danish case, provide complementary assets, but they can exploit those parts of the science base which are not reached by local firms. Doing so, they may strengthen the local cluster and be part of the process of increasing returns leading to a greater attractiveness to yet further investments. The regional policy issue is therefore to promote direct investment, primarily 'green field' investments, by marketing the unique competence of the region.

A second way to enlarge the actor base is to improve access to competent capital, that is, capital to which is attached technology or industry-specific competence, for university spin-offs. There need to be a large number of such actors involved. This is particularly true for science-based innovations which tend to go through a series of phases, each of which may be quite lengthy. The projects therefore need to be able to be 'traded' on the capital market.

Apart from stimulating the emergence of a local capital market, for instance organized by the universities or bridging institutions, the region needs to be linked up to the international capital market for the simple reason that small regions cannot be expected to host sufficient competent (specialized) capital. A regional policy issue is therefore to foster the development of a regional capital market and to facilitate the linkage of local actors to the international capital market.

To conclude, we have studied the renewal process in two, broadly speaking, medium-income regions within the European Union. We have found that the mechanisms leading to renewal have a strong regional dimension and, therefore, there is scope for a regional policy. The policy issues centre around the behaviour of the universities; the problem of necessarily having few and quite specialized local industrial actors; and how to enlarge the actor base, both by foreign direct investment and by improving access to the international capital market. Of course, the mechanisms identified are not unique to these regions, nor are the policy issues or the policy opportunities. What distinguishes one region from another is rather the way in which these are understood and handled by local policy actors, including the university leadership.

As far as EU regional policy initiatives are concerned the four case-studies did not facilitate any empirical comparison, given that Sweden only recently joined the EU. North Jutland has on the other hand received a substantial amount of support from EU industrial/regional policy funds for an extended period (for instance, from Target 2, 3, 4 and 5b and Stride programmes). The County Council of North Jutland has summarized a series of evaluations of the effects in terms of employment creation.[22] The total amount of public support during 1986–99 is estimated at around ECU 450 million, of which 210 million came from EU funds and 240 million from Danish government sources. The employment creation, including the effect of matching investments from the firms, has been calculated as roughly 13 000 jobs. No systematic industry-specific data are available, and it is not possible to give comprehensive statistical evidence for the two cases under consideration. However, several of the firms, the small ones as well as the largest, in the radiocommunications cluster (NorCOM) have received support from these EU funds. It is to be expected that the firms (in public) would be upwardly biased if asked about the importance of the money they have received. The present study of the NorCOM cluster has emphasized that the combination of entrepreneurial activity and university production of engineers and research has been decisive for the dynamic development of the cluster. The EU funds do not appear to have played a decisive role in that development. The support has apparently in many cases been applied as a supplement, which has increased the momentum of a development already under way. The consciousness about the specific systemic features of the cluster appears to have been one of the guides for reasonable use of the support.

Abbreviations used in this chapter

CPK	Centre for Personal Communications
DECT	Digital European Cordless Telecommunications
ETSI	European Telecommunications Standard Institute
GMDSS	Global Maritime Distress and Safety System
GPS	Global Positioning System
GSM	Global System for Mobile Communications
MIBA	Medical Informatics and Image Analysis
NMT	Nordic Mobile Telephone network
NOVI	Nordjyllands Videnpark
NUTS	Nomenclature of Statistical Territorial Units
OSO	Onsala Space Observatory
SMI	Centre for Sensory-Motor Interaction
VHF	Very High Frequencies (30–300 MHz)

NOTES

1. Since the publication of Freeman's (1987) and Hughes' (1983) books on the Japanese system of innovation and on large technical systems, respectively, there has been an explosion in the number of studies and approaches which share a 'systemic' view on the innovation process. Some focus on *national* innovation systems (e.g. Lundvall, 1992), whereas others study *sectoral* (Breschi and Malerba, 1997), *technological* (Carlsson and Stankiewicz, 1991) or

regional systems (e.g. Braczyk, Cooke and Heidenreich, 1998; Saxenian, 1994). Whereas most of the studies have been carried out within the broader area of economics of innovation, scholars in management have successfully incorporated firm strategy issues into a systemic view on innovation (Enright, 1995; Håkansson, 1987; Porter, 1990; Utterback and Afuah, 1995).

2. As is strongly emphasized in institutional economics (e.g. Edquist and Johnson, 1997), institutions greatly affect the specific path which a technology takes.
3. In the four case-studies, the delineation of systems varies somewhat. The cases in Western Sweden are essentially delineated on the basis of knowledge criteria whereas the North Jutland cases are rather built around industries. Of course, there is an overlap between the two ways of delineating technological systems and identifying the firms constituting industrial clusters, but they are not identical (see Holmén and Jacobsson, 1997, for a methodological discussion on this point). Each case-study incorporates a short discussion of how the system is delineated.
4. The industry categories are: traditional industries (ISIC 31, 32, 33, 36, 39), chemicals (ISIC 342, 35 (excl. 3522)), paper and pulp (ISIC 341), metal & machinery (ISIC 37, 381, 382 (excl. 3825)), 384 (excl. 3843)), automobiles (ISIC 3843), electronics (ISIC 3825, 383), pharmaceuticals (ISIC 3522) and instruments (ISIC 385).
5. This figure includes a broad range of disciplines in electronics, mathematics, physics and computer science.
6. From the more general innovation economics the division between Schumpeter Mark I and Mark II characteristics (Freeman and Soete, 1997; Verspagen, 1998) is an important analytical device to understand the coexistence of small development firms (Mark I) and the emergence of large-scale players (Mark II). This potential symbiosis between the rather dominant Mark I features of the cluster and the emergence of primarily foreign ownership-initiated Mark II features takes as its analytical point of departure Teece's (1986) discussion of (lack of) 'complementary assets'.
7. In 1993 Danish DECT Development was established at NOVI. This venture concentrated for two years on further specifications of DECT. Among the group of participants were two firms from North Jutland (Cortech and Lasat) and the Danish subsidiary of L.M. Ericsson. Parallel to Dancall's continued investments in cordless technology, part of it was divested to the German Deutsche Telefonwerke in Berlin (DeTeWe) and named CorTech. In 1997 Bosch moved the remaining part of the DECT activities in the newly acquired subsidiary to Germany.
8. Another small university spin-off, Partran, is based on research in acoustics with applications in the hearing aid industry. NOVI has supplied seed money for this venture, too.
9. Although the internal dynamics of the research process is often oriented very directly towards solving the specific needs of handicapped people.
10. This implies that, for instance, semiconductors are excluded. However, antenna and microwave technology has many relations to a broad range of applications such as wave guides, radio base stations, radar, satellite TV, mobile phones and microwave ovens.
11. However, the firm is also interested in the relations with the research performed at Chalmers – for instance through employing people with PhD degrees.
12. The regional economic importance of the firm is, however, of some size, given that it employs about 200 people.
13. The interests from the firms in their sourcing from the regional competence base are, in order of importance, the local pool of engineers (including the graduates), the PhDs, and the scientific research per se. Recently interest in the PhDs has increased considerably.
14. 'Intelligent antennas' give a much higher transfer capacity in mobile communication systems than traditional antenna systems. This is believed to be feasible since the intelligent antennas only 'look' in the direction of the users of the system and ignore everyone else.
15. Indeed, again by chance, the regional technological system is well equipped to deal with the expected technological discontinuity. In particular, the competence base in digital signal processing is strong at Chalmers.
16. Nobel Industries was at that time known as Bofors.

17. The parent company could provide capital and production facilities, but also had the required engineering skills, goodwill and marketing channels.
18. In cooperation with the Brånemark team, the next generation of dental implants will soon be ready for the market, giving possibilities for further expansion.
19. In the USA, a former Nobel Biocare director also set up a competing firm, and a number of the other start-ups were made by customers.
20. The potential market is patients with a defect in their middle ear, which is approximately 2 per cent of all hearing deficiencies. The P&B/Nobel product is the only implantable alternative on the market, but as the product is approximately three times as expensive as a traditional hearing device the diffusion has not taken off yet. Up to 1997, 5000 patients had been treated.
21. The development of the applications of bone implant technology discussed above spanned a fair number of years. Many times the development efforts certainly seemed fruitless, but Brånemark and other researchers continued against the odds, convincing the financiers that it was worth the investment. Moreover, when the first application was spun off to an industrial company the search for new applications continued with renewed strength.
22. The data are available at http://www.nja.dk/faelles/direk/erhvervs/resultat.html (in Danish).

REFERENCES

Audretsch, D.B. and M.P. Feldman (1994), 'Knowledge Spillovers and the Geography of Innovation and Production', paper presented at the Schumpeter Society Conference, Münster.

Braczyk, Hans-Joachim, Philip Cooke and Martin Heidenreich (eds) (1998), *Regional Innovation Systems. The Role of Governances in a Globalized World*, Reading: UCL Press.

Breschi, Stefano and Franco Malerba (1997), 'Sectoral Innovation Systems: Technological Regimes, Schumpeterian Dynamics, and Spatial Boundaries', in C. Edquist (ed.), *Systems of Innovation: Technologies, Institutions and Organizations*, London: Pinter.

Carlsson, Bo and Staffan Jacobsson (1991), 'Factory Automation and Government Policy', in B. Carlsson (ed.), *Technological Systems and Economic Performance: The Case of Factory Automation*, Amsterdam: Kluwer Academic Publishers.

Carlsson, Bo and Staffan Jacobsson (1997), 'Diversity Creation and Technological Systems: A Technology Policy Perspective', in C. Edquist (ed.), *Systems of Innovation: Technologies, Institutions and Organizations*, London: Pinter.

Carlsson, Bo and Richard Stankiewicz (1991), 'On the Nature, Function and Composition of Technological Systems', *Journal of Evolutionary Economics*, **1**(2), 93–118.

Dalum, Bent (1995), 'Local and Global Linkages: The Radiocommunications Cluster in Northern Denmark', *Journal of Industry Studies*, **2**(2), 89–109.

Dalum, Bent, Keld Laursen and Gert Villumsen (1997), 'Is There Such a Thing as a "Good" Export Specialization Pattern?: A European Perspective', paper presented at the EU-TSER/TEIS project seminar, Naples.

David, Paul A. (1988), 'Path-Dependence: Putting the Past into the Future of Economics', Technical Report, Institute for Mathematical Studies in the Social Sciences, Stanford University.

Edquist, Charles and Björn Johnson (1997), 'Institutions and Organizations in Systems of Innovation', in C. Edquist (ed.), *Systems of Innovation. Technologies, Institutions and Organizations*, London: Pinter.

Enright, M.J. (1995), 'Regional Clusters and Economics Development: A Research Agenda', paper presented at the Regional Clusters and Business Networks conference, Fredericton, Canada.

Freeman, Christopher (1987), *Technology Policy and Economic Performance – Lessons from Japan*, London: Pinter.

Freeman, Christopher and Luc Soete (1997), *The Economics of Industrial Innovation*, 3rd edn, London: Pinter.

Håkansson, H. (1987), *Industrial and Technological Development: A Network Approach*, London: Croom Helm.

Holmén, Magnus and Staffan Jacobsson (1997), 'A Method for Identifying Actors in a Knowledge Based Cluster', in M. Holmén (ed.), *Characterising and Analysing Regional Renewal: A Methodological Contribution*, Göteborg: Department of Industrial Dynamics, Chalmers University of Technology.

Hughes, Thomas P. (1983), *Networks of Power: Electrification in Western Society, 1890–1930*, Baltimore: Johns Hopkins University Press.

Jacobsson, Staffan and Annika Rickne (1996), 'New Technology-Based Firms and Industrial Renewal – an Analytical Framework', in A. Rickne (ed.), *New Technology-Based Firms and Industrial Renewal*, Gothenburg: Chalmers University of Technology, Department of Industrial Management and Economics.

Lundvall, Bengt-Åke (ed.) (1992), *National Systems of Innovation: Towards a Theory of Innovation and Active Learning*, London: Pinter.

Piore, Michael and Charles Sabel (1984), *The Second Industrial Divide*, New York: Basic Books.

Porter, Michael E. (1990), *The Competitive Advantage of Nations*, London: Macmillan.

Saemundsson, R., A. Rickne, Å. Lindholm-Dahlstrand and S. Jacobsson (1997), 'Regional Industrial Renewal: Technological Specialization and Employment in Swedish New Technology Based Firms', paper presented at the conference 'Industriell & Teknisk Utveckling', Fågelbrohus, Sweden, 26–28 November.

Saxenian, A. (1994), *Regional Advantage. Culture and Competition in Silicon Valley and Route 128*, Cambridge, MA: Harvard University Press.

Teece, David (1986), 'Profiting from Technological Innovation: Implications for Integration, Collaboration, Licensing and Public Policy', *Research Policy*, **15**(6), 285–305.

Utterback, J. and A. Afuah (1995), 'The Dynamic "Diamond": A Technological Innovation Perspective', Working Paper, MIT, Sloan School of Management, pp. 138–95.

Verspagen, Bart (1998), 'Large Firms and Knowledge Flows in the Dutch R&D System. A Case Study of Philips Electronics', mimeo, MERIT.

9. The implications of eastward enlargement for EU integration, convergence and competitiveness

Heather Grabbe, Kirsty Hughes and Michael Landesmann

1 INTRODUCTION

Eastward enlargement has major implications for EU political economy: in particular, it poses the unprecedented challenge of extending a single market over 25 or more economies of considerable heterogeneity. Enlarging successfully is likely to require attention to adapting EU policies to fit the new challenges raised by taking in up to ten Central and East European (CEE) countries. The economic effects of enlargement will be widespread and many are at present impossible to quantify with any certainty; however, it is important for the EU to consider the potential implications and what policies might be developed to deal with likely consequences. This process of policy review needs to start prior to the first accessions, given three factors: first, that integration is already occurring through trade, FDI and outsourcing; second, that some of the potentially adverse economic and social effects could affect the political economy of the enlargement process (including negotiations); third, that a new EU budget has to be negotiated in the next two years to run until 2006.

In considering the economic impact of enlargement, the relationship between *integration*, *convergence* and *competitiveness* deserves particular attention because emerging trends have implications that are likely to need policy responses. Enlargement has particularly important implications for these three issues because it will affect them all.

This chapter documents and discusses existing disparities and integration between the EU and CEE regions. Section 2 gives an overview of the current range of economic performance and income across the EU-15 and the CEE-10. Sections 3 and 4 look at trends in trade and FDI flows between the EU-15 and CEE-10 since 1989, focusing on trade specialization and geographical patterns of integration and drawing out some of the implications for competitiveness and accession for the CEE region.

2 ECONOMIC DISPARITIES BETWEEN CEE AND THE EU

All of the CEE-10 countries are poorer than the poorest EU member states, and average real GDP per head (on a PPP basis) in CEE is only 32 per cent of the average for the EU-15. However, the range of economic performance and income across CEE is considerable. Looking at income first, the Czech Republic, Hungary, Poland, Slovakia and Slovenia are between one-third and two-thirds of the EU average, whereas the rest are below one-third of the EU average in 1995 (see Table 9.1). The differential between the richest and poorest CEE countries is considerable: Latvian GDP per capita is only one-third the size of that in Slovenia, to take the extremes.

Table 9.1 GDP per head on PPP basis, 1995

	ECU per head	Percentage of EU average
Slovenia	10.110	59
Czech Republic	9.410	55
Slovakia	7.120	41
Hungary	6.310	37
Poland	5.320	31
Bulgaria	4.210	24
Lithuania	4.130	24
Romania	4.060	23
Estonia	3.920	23
Latvia	3.160	18
CEE-10	*5.530*	*32*
Luxembourg	29.140	169
Denmark	19.960	116
Belgium	19.340	112
Austria	19.320	112
Germany	19.070	110
France	18.520	107
Netherlands	18.390	107
Italy	17.770	103
Sweden	17.390	101
United Kingdom	16.580	96
Finland	16.550	96
Spain	13.230	77
Portugal	11.620	67
Greece	11.320	66
EU-15	*17.260*	*100*

Source: CEC, 1997 (Agenda 2000).

However, the disparities presented by the CEE applicants are not massively greater than those already existing within the EU. The gap in 1995 between the average income level of the CEE countries and the poorest two EU members (Greece and Portugal) is only half the size of the differential between Greece and Portugal and the richest two EU members (Luxembourg and Denmark). In taking in the four richest CEE countries, the EU would be extending its range of income disparities by a further 25 percentage points.

In Tables 9.2a and 9.2b we present some additional information on wage rates, productivity levels and unit labour costs in manufacturing of a range of CEE economies in relation to Austrian levels (Austria = 100). We can see that CEE wage rates are only a fraction of West European (Austrian) wage rates when evaluated at current exchange rates (Table 9.2a). Comparisons of wage rates (and labour unit costs) at current exchange rates are affected by the strong undervaluation of many of the CEE exchange rates.[1] In Table 9.2b we also present, therefore, wage rate and labour unit cost comparisons calculated

Table 9.2a Wage rates, productivity and unit labour costs (ECU), total manufacturing (Austria 1993 = 100)

		1991	1992	1993	1994	1995	1996
Czech Republic	Wages	4.9	5.8	7.8	9.1	10.6	12.4
	Productivity	42.1	44.1	43.4	45.6	50.8	55.0
	Unit lab. costs	11.5	13.2	17.9	19.8	20.8	22.4
Hungary	Wages	8.2	9.5	11.2	11.9	11.1	
	Productivity		25.7	30.4	36.7	40.9	
	Unit lab. costs		36.9	36.9	32.6	27.2	
Slovak Republic	Wages	4.9	5.4	6.7	7.5	8.6	9.9
	Productivity	44.5	42.9	39.0	41.9	45.2	46.8
	Unit lab. costs	11.0	12.7	17.2	18.0	19.1	21.1
Poland	Wages	5.8	7.0	7.9	8.6	10.2	12.1
	Productivity		31.1	35.5	40.5	44.4	49.6
	Unit lab. costs		22.4	22.4	21.2	23.0	24.3
Romania	Wages	3.4	2.2	2.9	3.1	3.6	
	Productivity	19.9	17.5	18.9	22.6	27.8	
	Unit lab. costs	17.2	12.3	15.4	13.6	12.9	
Slovenia	Wages	19.8	18.9	21.7	23.9	27.8	28.7
	Productivity	37.9	35.9	37.9	42.5	46.0	48.8
	Unit lab. costs	52.4	52.6	57.1	56.2	60.5	58.8

Source: WIIW – Structural Database, derived from national industry statistics.

from purchasing power parity (PPP) rates. The gaps are then somewhat reduced but still remain very large. Comparisons of (labour) productivity levels are notoriously difficult, as they require a comparison of 'real' output levels across countries at a set of base year price levels (in our case 1993). The comparisons (using GDP PPP rates) show that the gaps in wage rates are greater than in labour productivity which, of course, means that unit labour costs are substantially below Austrian levels; or, expressed differently, the gaps in productivity levels are more than compensated by gaps in wage rates, so that CEE countries still remain very competitive in unit labour cost terms. The most significant further gap which affects CEE's competitiveness is the gap in product qualities (to be interpreted in a wide sense, including brand name, product design, packaging, marketing, distributive channels, etc.) a subject which is discussed in Chapter 3 of this book (see Jansen and Landesmann, Ch. 3).

Table 9.2b Wage rates, productivity and unit labour costs (PPP basis), total manufacturing (Austria 1993 = 100)

		1991	1992	1993	1994	1995	1996
Czech Republic	Wages	22.0	23.6	25.9	27.8	30.4	33.4
	Productivity	42.1	44.1	43.4	45.6	50.8	55.0
	Unit lab. costs	52.2	53.5	59.6	60.9	59.8	60.7
Hungary	Wages	18.3	20.0	21.3	22.6	22.6	
	Productivity		25.7	30.4	36.7	40.9	
	Unit lab. costs		78.0	70.1	61.7	55.2	
Slovak Republic	Wages	19.5	20.5	22.3	23.8	26.0	29.0
	Productivity	44.5	42.9	39.0	41.9	45.2	46.8
	Unit lab. costs	43.9	47.8	57.2	56.9	57.7	61.9
Poland	Wages	14.9	15.5	14.5	15.8	17.6	21.0
	Productivity		31.1	35.5	40.5	44.4	49.6
	Unit lab. costs		49.8	40.8	39.1	39.7	42.3
Romania	Wages	12.6	11.7	11.1	11.1	13.3	
	Productivity	19.9	17.5	18.9	22.6	27.8	
	Unit lab. costs	63.3	66.7	58.7	49.3	47.8	
Slovenia	Wages	0.0	11.7	25.9	31.3	35.6	38.9
	Productivity	37.9	35.9	37.9	42.5	45.9	48.8
	Unit lab. costs	0.0	32.5	68.4	73.8	77.4	79.8

Note: For CEECs, PPP rates of current years were used, for Austria the PPP rate for 1993.

2.1 Prospects for Catching Up with EU Income Levels

The importance of economic disparities between new and old member states will depend on how quickly they narrow prior to and after accession. This factor is in turn contingent on economic performance, and particularly on GDP growth rates.

Overall economic performance has been closely watched as one indicator of when each of the CEE applicants might be considered ready to join the EU. The CEE economies began transition with differing circumstances, and there is still considerable (and to some extent growing) heterogeneity in their performance. Table 9.3 shows GDP growth.

Table 9.3 Development of real GDP in Eastern Europe 1990–97, real change in per cent against preceding year

	1989	1990	1991	1992	1993	1994	1995	1996	1997	1993–97 average
Czech Republic	4.5	–1.2	–11.5	–3.3	0.6	3.2	6.4	3.9	1.0	3.0
Hungary	0.7	–3.5	–11.9	–3.1	–0.6	2.9	1.5	1.3	4.4	1.9
Poland	0.2	–11.6	–7.0	2.6	3.8	5.2	7.0	6.1	6.9	5.8
Slovak Republic	1.0	–2.5	–14.5	–6.5	–3.9	5.0	7.3	6.6	6.5	4.2
Slovenia	–1.8	–4.7	–8.9	–5.5	2.8	5.3	4.1	3.1	3.8	3.8
CEEC (5)	1.2	–6.8	–9.5	–0.8	1.7	4.4	5.8	4.7	4.9	4.3
Bulgaria	–1.9	–9.1	–11.7	–7.3	–1.5	1.8	2.1	–10.9	–6.9	–3.2
Romania	–5.8	–5.6	–12.9	–8.7	1.5	3.9	7.1	4.1	–6.6	1.9
CEEC (7)	–0.2	–6.8	–10.3	–2.6	1.4	4.1	5.8	3.5	2.1	3.4
Croatia	—	–7.1	–21.1	–11.7	–8.0	5.9	6.8	6.0	6.5	3.3
Russia	—	–3.0	–5.0	–14.5	–8.7	–12.7	–4.1	–4.9	0.4	–6.1
Ukraine	—	–2.6	–11.6	–13.7	–14.2	–23.0	–11.8	–10.0	–3.2	–12.7

Source: WIIW statistics.

As we can see, GDP growth is still vulnerable to strong fluctuations especially in the 'eastern' CEECs in which the implementation of the major reform programmes is far from complete (Bulgaria, Romania, Russia, Ukraine); in those CEECs in which growth has remained positive, there is still the possibility of temporary set-backs, usually induced by exchange rate and balance of payments crises (Hungary 1994/5, Czech Republic 1997), but there are signs indicating that most of the 'western' CEECs (Poland, Hungary, Czech Rep., Slovakia, Slovenia) are embarking upon a catching-up growth path with growth rates significantly above the EU rates. Thus, over the period 1994–98 the European Commission forecasts an average real GDP growth rate of above 4 per cent, in comparison with about 2.5 per cent for the EU (Hallet, 1997). As the CEECs have embarked upon their catching-up processes (implying higher growth rates than their main trading partners) trade balances have continued to deteriorate in all of the countries, particularly with the EU, their main trading partner (see also section 3). External debt stock has, however, been progressively reduced.

The most recent studies (for instance EBRD, 1995 and 1997) conclude from quantitative predictions that the transition economies would be generally well placed for rapid growth after the initial period of transition because of their high level of acquired skills and the large initial shortfall in productivity; over the long term, the transition economies could thus benefit from rapid productivity growth owing to the combination of a well-educated population and relatively low initial productivity levels. These factors point to potential long-term average annual growth rates in the range of 4–7 per cent for most countries.

It could thus take several decades for even the most advanced CEE countries to reach average EU income levels. The World Bank estimates that at present growth rates the Visegrad countries and Slovenia are 20 years away from average EU incomes in 1994. Accession itself will most likely affect rates of catch-up across CEE, as some are brought in while others still wait outside. Sachs and Warner (1996) argue that growth levels would be boosted by early EU membership because it would lock in reforms, provide access to the Single Market and encourage investor confidence.

Given the outlook for catch-up and for integration, the EU faces the prospect of having to deal with considerable disparities between member states' income levels, and for a long time period. The range of income differences is wide overall, although the most advanced CEE countries are not poorer than previous applicants were. Overall, the challenge for the EU lies in devising policies for dealing with the inequalities in income across a new EU being formed over a considerable period of years. Moreover, these disparities are not just between old and new member states, but also between the CEE economies. There are regional inequalities within the applicant states as well in terms of income levels and unemployment rates (see Hallet, 1997).

3 INTEGRATION SO FAR: TRADE AND FDI

Integration between the EU and CEE countries brings the potential to reduce inequalities of income and economic performance across the region; however, the effects may not be evenly distributed. The following section examines integration through trade and FDI since 1989, showing the geographical patterns that are emerging, and then discusses likely future trends and their potential implications for competitiveness.

3.1 Trade

Trade growth between the two regions has been very dynamic since 1989. As is shown in Table 9.4, exports to the CEE-6[2] from the EU grew by an average of 25 per cent per year in the first seven years of transition, and imports from CEE countries by 21 per cent, although they fluctuated considerably from year to year. Factors behind these fluctuations include the recession of the early 1990s in western Europe, the plunge and then gradual recovery of output in CEE after the initial shock of transition, periodic currency under- and over-valuations (in relation to external balances) and steps in trade liberalization on both sides.

Table 9.4 Growth in trade since 1989

Annual rate of growth over previous year (%)	1989	1990	1991	1992	1993	1994	1995
EU-12 exports to CEE-6	16	47	16	28	9	19	38
1995/89 % increase	131						
EU-12 imports from CEE-6	8	30	16	22	–3	31	40
1995/89 % increase	185						

Source: Own calculations based on IMF (1996); calculated from current dollar price data.

A massive reorientation of CEE trade took place after 1989. By 1995, the EU had become by far the most important market for CEE exports, accounting for over half of the total. In part, the rapidity of the reorientation is accounted for by the low base from which EU–CEE trade grew. The Europe Agreements also played a role by providing a framework for continued liberalization and locking both sides into a time frame for reducing trade barriers.

Economic interpenetration between the two regions is asymmetric, with the EU having a lesser economic interest in CEE than vice versa. For the CEE region, trade with the EU-15 is more important than trade with any other region, including intra-CEE trade. The EU accounts for more than half of total imports into the region, and it accounts for two-thirds of imports for five of the CEE-10. However, degrees of trade dependence on the EU vary across the CEE-10.

Despite this growth and reorientation, however, trade with CEE countries still accounts for less than 10 per cent of total extra-EU trade, primarily because these economies are still small in comparison with the EU. However, the region takes a greater share of EU exports than does Japan, and trade is still growing rapidly. Thus, while the CEECs' combined share in the EU's exports was about 8 per cent in 1996, its share in the growth of the EU's exports over the period

1993–96 was about 15 per cent. Moreover, significant further expansion is likely in line with growth in GDP and incomes in CEE, whereas many of the EU's other markets are at or are reaching maturity. From this point of view, the CEE countries represent an important market with considerable potential for the EU.

The diversity in the shares taken by the different EU members in trade with the CEE region is shown in Table 9.5. Germany accounted for by far the largest share of CEE trade among EU countries in 1989, although this was in the context of very low levels of overall EU trade with the region. Despite the strong growth in overall trade flows and the change in economic relations with CEE, Germany still dominates trade, accounting for half of total EU exports to the region.

Table 9.5 Shares of main EU exporters

% of EU-15 exports to CEE-6	1989	1995
Germany	32.83	47.68
Italy	14.14	10.89
Austria	8.92	7.30
France	8.95	5.41
United Kingdom	7.09	5.07
Netherlands	5.02	3.45
Other	23.04	20.20
Total	100.00	100.00

Source: Own calculations based on IMF (1996).

The overall trade balance is a large net surplus for the EU, despite the asymmetry of the Europe Agreements in lifting restrictions on CEE exports first. Even though nearly all restrictions on trade in industrial goods have now been removed, the trade imbalance has continued to grow. One factor seems to be CEE imports of capital and intermediate goods to replace capital stock and underpin restructuring during transition, although there has also been pent-up demand for western consumer goods and generally a high propensity to import. Varying levels of competitiveness in various sectors in the CEE countries also play a role, as do exchange rate developments (partly reflecting policy and partly international capital movements) which led to 'over-valuations' of some of the CEECs' currencies from the point of view of external balances. Finally, catching-up processes (involving higher growth rates than one's principal trading partners) are usually accompanied by current account imbalances (see Landesmann and Poeschl, 1997).

Behind the aggregate EU surplus lie important differences in the trade performance of individual EU members. Most EU countries show a surplus in trade with most CEE countries; the exceptions are Portugal, Greece, Spain and Denmark. The relatively worse trade performance of these countries and the small size of their overall trade with CEE could have an effect on perceptions of the economic costs and benefits of enlargement; however, national political debate on this issue has so far been overshadowed by other concerns about enlargement, including its potential budgetary impact.

In summary, the geographical reorientation of trade towards the EU since 1989 has been rapid, as has overall growth, and Germany now accounts for half of all EU exports to and imports from CEE. We shall now move on to report on the distinct *pattern of specialization* which characterizes trade relations between the EU and CEE.

The rapid and deep liberalization of external relations after 1989 has led to a rapid emergence of a *new pattern of trade specialization* in EU–CEE trade which is mostly in conformity with predictions made by traditional trade theory. In Appendix 9.1 we show by means of a factor contents analysis that the CEE economies have strong deficits in skill- and technology-intensive branches and that there is some positive specialization towards labour-intensive branches (see also the evidence in *European Economy*, 1995 and Landesmann, 1995). However, while there is evidence for industrial and trade specialization in the Central and East European countries in conformity with comparative levels of economic development, factor endowments, etc., the emerging trade structures also reveal the impact of the *embarkation of some of the CEECs on a process of catching up* with the more advanced western European economies (in technological and product quality terms).

Thus, two forces have been operating alongside each other: (a) the accentuation of patterns of inter-industry specialization, following the strong liberalization of trade, and (b) the beginnings of a process of catching up (very differentiated across the different CEECs) which is traditionally associated with a decline in strong patterns of inter-industry specialization and an increase in intra-industry trade (for details on this argument and evidence, see Landesmann, 1995).

Evidence on catching up suggests the emergence of strong *heterogeneity across the CEE economies*. For certain economies, notably Hungary, the Czech Republic and Slovenia, we observe strong increases in intra-industry trade with EU trading partners (see Table 9.6) and a decline in the strong comparative bias against skill, R&D and, lately, capital-intensive branches (see Table 9A.1 in Appendix 9.1). For other countries such as Romania and Bulgaria – with Poland occupying a middle position – patterns of inter-industry specialization compatible with differential factor endowment positions between eastern and western European economies are further reinforced. Comparisons of both industrial and trade structures of CEE countries with northern and southern European economies show that the more advanced of the CEE economies occupy a middle position between the industrially more advanced northern EU and ex-EFTA countries on the one hand and the southern European economies (Spain, Portugal, Greece) on the other (see Urban, 1997; Landesmann, 1998).

Table 9.6 The Grubel–Lloyd indicators of CEECs', Austria's, and EU-external total manufacturing trade with EU-12

	1989	1990	1991	1992	1993	1994	1995	1996
Hungary	0.483	0.498	0.519	0.532	0.546	0.581	0.600	0.617
Poland	0.418	0.389	0.407	0.429	0.439	0.436	0.473	0.465
Czech Republic					0.604	0.618	0.649	0.669
Slovak Republic					0.433	0.472	0.531	0.667
Slovenia					0.575	0.636	0.654	0.667
Croatia							0.459	0.443
Bulgaria	0.397	0.420	0.388	0.391	0.447	0.431	0.402	0.409
Romania	0.291	0.298	0.339	0.293	0.293	0.309	0.342	0.339
Russia							0.172	0.165
Ukraine							0.223	0.267
Austria	0.734	0.742	0.751	0.758	0.747	0.753	0.707	0.715
EU-external	0.693	0.693	0.706	0.704	0.693	0.702	0.688	0.687

Note: The GL indicators were calculated on trade flows across 108 manufacturing (NACE 3-digit) branches using the formula $GL = 1 - \Sigma_x \mid E_x - M_x \mid / \Sigma_x (E_x + M_x)$ where E_x and M_x stand respectively for exports and imports in branch x.

Hence the analysis of trade relationships between the CEECs and the EU presented in Appendix 9.1 reveals significant inter-industry specialization in CEE's trade relationships with the EU. The specialization patterns have been affected (partly strengthened, partly changed) by developments since 1989. The more western CEE economies did not experience a reinforcement of this pattern of specialization and, in the cases of Slovenia, Hungary and the Czech Republic, we can perceive a significant closure of the gap in skill- and R&D-intensive areas; in the cases of Poland, Romania and Bulgaria the gap remains high or is even widening. This process of differentiation reflects differences both in the developmental levels and in the dynamics of industrial change that has taken place in the various CEE economies since 1989. This process of differentiation across CEE is also clearly evident in the study on 'quality differentiation' presented in Chapter 3.

3.2 Foreign Direct Investment

FDI inflows into Central and Eastern Europe were relatively low at the start of transition, but they have risen substantially subsequently (see Estrin, Hughes and Todd, 1997 for a more detailed discussion). Inflows into the CEE-10 countries account for most of the inflows into transition economies at more than 80 per cent, with the other transition economies, including Russia and the CIS, accounting for about 20 per cent. Among the CEE-10, Hungary, Poland and the Czech Republic have received the bulk of the inflows, with over two-thirds of the total. Data on FDI in Central and Eastern Europe have not been entirely reliable, and in the early years of transition there were discrepancies between international data sources and national data; in particular, there were problems with investments by multinationals being recorded as being from their foreign subsidiaries. There is more convergence in recent data, however, and the UN now estimates cumulative inflows of 31.5 billion dollars between 1989 and 1995 for the whole region.

Table 9.7 sets out the UN data for the CEE-10 from 1990 to 1996. Owing to a surge in inflows in 1996, Poland has received nearly one-third of investment inflows followed by Hungary, the Czech Republic and then Russia. Hungary's

Table 9.7 Cumulative FDI inflows to CEE and CIS, 1990–96[a]

	Cumulative flows (US$m)	% of total flows	Inflows per capita (US$)
Poland	13503	28.7	350.73
Hungary	12936	27.5	1255.92
Czech Republic[b]	6938	14.8	673.61
Romania	1595	3.4	70.26
Slovakia[b]	1007	2.1	189.96
Estonia	799	1.7	532.67
Latvia	760	1.6	304.00
Slovenia	688	1.5	344.00
Bulgaria	502	1.1	59.76
Lithuania	296	0.6	80.00
CEE-10 total	*39024*	*83.0*	*370.95*
Russian Federation	4054	8.6	27.34
Other CIS and CEE countries[c]	3949	8.4	26.83
Total flows to CEE and CIS	47027	100.0	117.36

Notes:
a 1996 figures are UN estimates.
b Own calculation made of allocation of FDI inflows to Czechoslovakia prior to 1993.
c Includes the other CIS countries and Albania.

Source: Own calculations based on UN (1997) and World Bank (1996).

inflows are particularly large relative to its size, while inflows into Russia are small. These four countries take around 80 per cent of all FDI into the region, showing the concentration of flows on a few destinations. While market size is one important determinant of FDI flows, success in political and economic transition is also a key factor (Estrin et al., 1997). Thus success breeds success; stable, successful transition attracts FDI, which further contributes to transition. Moreover, investors show the same preferences as the EU and NATO in terms of the countries' relative progress.

The western source countries for FDI into CEE vary. Germany and the United States dominate with about 20 per cent of total investment each (Estrin et al., 1997). France has about 7 per cent and the UK about 4 per cent. As with the trade data, the FDI data reinforce the picture of unevenness in EU–CEE integration, with Germany playing a particularly important role and the UK, notably, a much smaller role in comparison with their respective shares of international trade and investment. However, there is considerable variation in Germany's shares of FDI across the countries and overall Germany is not as dominant as in the trade flows. Shares of FDI are thus fairly similar to relative shares of trade among the EU member states, although there is considerable variation by year because of the effect that large single investments have on the total.

A number of commentators and analysts have expressed disappointment and concern at low levels of foreign direct investment into CEE (World Bank, 1996; Baldwin, 1994), but it is not clear that the data support this conclusion (see Estrin et al., 1997). FDI flows were low initially but this is inevitable given the risks and uncertainties of transition and the need to develop appropriate legal and institutional frameworks and market structures. Subsequently, FDI growth rates have been very high indeed and per capita levels into the top three CEE countries compare well with those into a number of EU member states and also with flows into some of the main developing country recipients of FDI (Estrin et al., 1997). Hungarian FDI per head places it just after France, Spain and the UK (the biggest EU recipients) and ahead of Portugal. On a per capita basis, flows into the leading three CEE countries are substantially higher than those into the leading host developing countries, although, in absolute terms, these countries mostly (though not all) receive more (Estrin et al., 1997). There is by now detailed evidence that FDI in CEECs contributes substantially to their corporate and industrial restructuring; firms with foreign ownership (FIEs) show significantly higher investment rates, higher export intensity and higher productivity levels (see Hunya, 1997; Zemplinerova, 1997). There is also evidence of extensive outsourcing activity and OPT (outward processing trade) in particular branches (see, for instance, Graziani, 1997).

It is clear, however, that FDI flows into CEE have been concentrated to a small number of countries. Important factors here include success in economic and political transition, market size, labour costs, location and government policy (Estrin et al., 1997). While market size is important and differs across countries, the importance of success in transition suggests that other countries will begin to attract substantially more FDI as they proceed with economic development. Overall, it is unclear how important accession to the EU will be in affecting future investment flows. Given that flows are high in Hungary, the Czech Republic and Poland, substantial investment both for market and labour cost reasons is likely to continue to occur prior to EU accession. Expectations of future EU membership may influence investment (as happened with previous enlargements of the EU), but flows also tend to correlate with success in economic and political transition. The high rates of increase in FDI anticipated by some (for example, Baldwin, François and Portes, 1997) after EU accession may not occur if substantial investment flows are actually experienced beforehand.

4 FUTURE TRENDS IN INTEGRATION

The EU–CEE economic integration seems likely to continue its rapid pace, given the high levels of trade growth since 1989 and the increasing attraction of FDI into CEE discussed in this section. Estimates made of the potential for future growth in EU–CEE trade range widely and are based on a variety of approaches and models (see Collins and Rodrik, 1991; Wang and Winters, 1991; Hamilton and Winters, 1992; Hughes and Hare, 1992; Baldwin, 1994; Brown et al., 1997; Neven, 1995 for early estimates). So far trade has grown in line with the general prediction made by various authors that trade could increase 3–4 fold (Hamilton and Winters, 1992; Baldwin, 1994, 1995a; CEC, 1994; Cadot, Faini and de Melo, 1995). In the long term, it could continue to rise at relatively high rates, that is, exceeding growth in overall EU trade as potential dynamic gains from rapid transfer of technology and managerial skills are realized and a virtuous circle of trade expansion and faster growth develops (Neven, 1995). Much will depend on the path of economic transition in CEE and other factors, such as exchange rate policy and trends in foreign investment.

There is much scope for Western firms (EU, but also non-EU) to make use of the substantial increase in the spread of production conditions (productivity levels, wage rates, skills) across the European continent. In some of the CEECs substantial segments of industrial production could be integrated into international production chains. Attention will increasingly focus on the extent of 'spillover effects' between the activities with foreign corporate participation and the rest of the domestic economies. Evidence from Asian experience suggests that the depth and breadth of these spillovers are vital for the overall development process. Corporate strategies in the current era are designed to exploit vertically differentiated production conditions globally (see Feenstra and Hanson, 1995a, 1995b; Markusen et al., 1996). A variety of studies (see, for instance, Borrus, 1995; Doherty, 1995) have pointed out that, in order to organize their operations worldwide, US and Japanese firms, in particular, have built up corporate cross-border networks (sometimes with and sometimes without ownership control) which have exploited the differentiation in technological capabilities and cost conditions across South East Asian countries and regions. As the degree of differentiation of production conditions has vastly increased in Europe as a result of East–West European integration, it is quite likely that similar networks initiated by western European (but also US, Japanese and Korean) firms will evolve as part of the overall economic integration process of western and eastern Europe.

The *relationship between trade integration, FDI flows and labour market developments* has occupied many economists in the West over the past few years (for an overview of this – largely American – literature, see, for instance, Baldwin, 1995b; for a review of the debate see Wood, 1995; Lawrence, 1996). Careful studies (see, for instance, Murphy and Welch, 1991; Borjas, Freeman and Katz, 1992; Leamer, 1994; Sachs and Shatz, 1994) found significant effects of evolving international trade patterns, FDI and migration flows on employment and wage structures in the West although there is still substantial disagreement about the size of these effects. These studies mostly concern developments in the 1980s and concentrate on North–South trade and migration patterns. It is clear that this topic is of great relevance for East–West European relationships, particularly for countries in close geographical proximity.

However, there is a danger in applying standard trade theory in too simplistic a manner to the situation of increasing trade and corporate links between catching-up and more advanced economies. The evidence of strong vertical product and production differentiation between eastern and western European producers implies that some of the conjectures of traditional trade theory – Heckscher–Ohlin–Samuelson (HOS) theory – should be applied in a very cautious manner to draw out the implications of the fast-moving process of East–West European integration. In particular, the Stolper–Samuelson theorem which conjectures that a strong global pressure towards factor price equalization would be exerted through competition in the product markets, thereby strongly affecting relative and real factor earnings, has to be applied in a manner which takes full account of the degree of vertical product differentiation amongst trading partners. Catching-up economies, particularly those with very strong gaps in capability structures, are operating in different technological and organizational environments and, furthermore, are catering largely to different 'quality segments' of international product markets than are the more advanced western European producers. Nevertheless, we would expect the type of dynamics analysed in the North–South trade and FDI literature to gain momentum over time, as eastern European producers gain weight in western European markets and in intra-corporate European production chains; as the 'quality segmentation' in product markets weakens and, hence, as East European producers (or subsidiaries of western companies in eastern Europe) start to compete directly with western producers over a wider range of products in western and home markets.

Theory suggests that the strongest pressure of adjustment in the West would be exerted by the emergence of a strong pattern of inter-industry specialization; we feel this, however, will not be characteristic for the more advanced of the eastern European economies, given the scope for catching up in these economies. There, the already growing tendencies for intra-industry trade will tend to build up and strengthen the pressures for a quality (and skill) upgrading process on producers in the western economies. In this context we should emphasize that *real wage catch-up* is as much a feature of an overall catching-up process and, hence, what are now considered as extraordinarily wide wage gaps between eastern and western Europe will get eroded over time. The competitive challenge is a function of the relationship between 'cost' and 'quality' catching up, and here we expect that eastern European producers are going to differ from the East Asian 'miracle economies' in that these latter economies managed to mount a strong challenge to western producers by allowing, for considerable periods, quality catching up to outstrip cost catching up; this is less likely to be the case in eastern Europe where social aspirations and political possibilities tend to exert a stronger pressure towards rapid real income growth.

The CEE countries are already gaining many benefits from economic integration prior to actually becoming members of the EU. In addition to gaining foreign investment, preferential access to EU markets and increased competition, the CEE countries are being locked into a framework for economic liberalization and adjustment to EU norms through the pre-accession strategy which itself fosters economic integration. Integration may also have had some costs, such as overly rapid market opening and a possibly artificial bias towards EU markets rather than intra-CEE integration (Drábek and Smith, 1995). Attempting to follow EU policy models may also have costs if other institutional structures are more appropriate. Furthermore, undergoing a process of transition to a market economy at the same time as orienting trade towards the EU may make CEE countries particularly subject to the influence of EU integration in encouraging the development of some sectors and the decline of others.

4.1 The Impact of Accession

Once accession actually takes place, how much difference will it make to integration? For the CEE countries, there could be significant benefits in several areas. The first is that accession would also give the CEE countries a say in EU decision-making. This issue was central to EFTA members' motivations for joining the EU. Second, there are several blocked areas which are unlikely to see much progress before accession. Particularly important are liberalization of trade in agricultural goods and lifting of contingent protection provisions, both of which would occur once CEE countries were full members.

The third area is access to the Single Market, which is generally expected to give a considerable boost to CEE exporters, as well as further stabilizing CEE economic policy. The likely effect is difficult to assess because of the lack of consensus about the impact of the Single Market Programme on the EU-12 countries, particularly because of the need to take into account dynamic effects and scale effects. Moreover, the timetable for extension of the Single Market is uncertain, and there are likely to be transitional periods, derogations and extension of only some of the four freedoms to begin with.

The Single Market seems likely to have a lesser effect on CEE economic development than it did on the EU-12, whose trade had already reached a steady state by the mid-1980s; nevertheless, additional trade as a result of access to the Single Market may have a relatively small impact on the CEE countries in comparison with the strong effects of the general opening of trade since 1989. What needs much more analysis is the economic impact on CEE of taking on the Single Market *acquis*.

For the EU, the economic impact of enlargement is likely to be smaller, given the small size of the CEE economies. Studies published by CEPR (CEPR, 1990; Baldwin, 1994; contributions to Faini and Portes, 1995; Winters, 1995; Baldwin et al., 1997) and others (for example Brown et al., 1997) suggest that the aggregate impact will be small and mostly positive. The most recent estimates (Baldwin et al., 1997) suggest an overall gain for the EU-15 of about 10 billion ECU in real income. However, Baldwin et al. find that these gains are likely to be very unevenly distributed across the EU countries. The trade data presented above suggest that beyond the three main traders (Germany, Austria and Italy) there is currently a relative lack of economic interest in the CEE countries in comparison with interests in the EU and third countries. Whether this interest increases depends on whether the rest of the EU can catch up with Germany's level of integration. It is questionable to assume that Germany's trade and investment levels will reach a steady state in relation to other EU economies. Germany has already pulled further ahead of the rest of the EU, and German exporters and investors could continue to build on any first-mover advantage in terms of market knowledge, brand-building and FDI to increase their market share.

With regard to the sectoral distribution of gains and losses, Faini and Portes (1995) conclude from a number of studies that the impact of rapid import growth would be negligible and that trade with southern Europe is likely to be mostly intra-industry, with few distributional implications. They and other authors (Rollo and Smith, 1993; Dittus and Andersen, 1995) argue that even sensitive sectors are unlikely to suffer large adjustment costs. Wyplosz (1995) concludes that although there are strong effects on wages and employment in specific industries, the results are not systematic across industries and countries. Particularly important in the impact of enlargement on specific industrial sectors could be the likelihood of a high proportion of intra-industry trade, which would lower transaction costs (see Faini and Portes, 1995; Drábek and Smith, 1995). From the available evidence it does look as if the longer-term comparative advantage of some of the more advanced of the CEE countries might not necessarily be cheap labour per se but a *relatively cheap skilled labour force*, although due account has to be taken of important gaps in the skill structures. Exploitation of this potential presumes that necessary structural skill adjustment processes proceed successfully (supported by suitable government schemes of training and retraining and an overhaul of educational and training structures) and without too much effect of high transitional unemployment on skill erosion. The current emphasis of FDI on more capital- and technology-intensive activities in manufacturing (which are also areas in which the complementarities between capital and skilled labour are particularly high) in some of the CEE countries can be taken as evidence that their longer-term comparative advantages do not seem to lie exclusively in low-tech, low-skill production.

4.2 Implications for Competitiveness

Economic integration with CEE countries could affect the competitiveness of West European industries if the region proves to be an area of dynamic growth and low production costs. The region could cause structural changes in West European industry if cheaper CEE inputs and market opportunities for EU firms to expand in the CEE countries increase the competitiveness of the parts of western Europe most integrated with CEE relative to the rest of the region (Hughes, 1996).

Overall, the continued importance of German firms in CEE markets could cause the German economy to benefit most among EU countries from the potential structural changes in West European industry which result from integration with CEE economies. German industry could see some activities continuing to relocate to and expand in CEE locations, and a general refocusing on higher value-added activities. Such an outcome would increase Germany's competitiveness relative to the rest of the EU. Given recent concerns about German competitiveness, this is to be welcomed, but rapid changes in relative economic performance across the EU could cause tensions, particularly in the context of EMU.

Integration depends principally on the size of trade, OPT and FDI flows, how and why they vary by country (as discussed above), and the relative comparative advantage and competitiveness of the CEE countries *vis-à-vis* the EU. This last factor is not static, and it is unclear how it will change over time. Competitiveness relates both to inter-industry trade and intra-industry trade, the latter being particularly associated with catching up; in fact, so far intra-industry trade seems to be more CEE trade in low-quality goods and the EU in high-quality goods. The CEE countries might have been developing a comparative advantage in cheaper labour and high levels of skills; this should lead to a beneficial restructuring of the European economy overall, but it may have a negative effect on particular sectors and groups in the EU.

As mentioned above, the analysis of the impact of evolving industrial specialization patterns between eastern and western Europe (involving both trade relationships and direct corporate integration) upon the *labour markets* in both parts of Europe is an extremely important topic and still very much under-researched. For some time to come, competition from CEE producers will exert significant pressure upon the lower-cost, lower-quality segments of West European production. This will contribute towards additional pressure on industrial and skill upgrading, particularly in the countries which are geographically more exposed to such competition. From the point of view of longer-term industrial development, such a move – if successfully managed – should have a positive impact on 'endogenous' growth in the more advanced western European economies. Government policies in the West should be directed to support the necessary skill and technological upgrading process.

Another sensitive issue concerns *migration of labour*. The interdependence between upward movements in the sophistication of the industrial structure and the demand for skilled labour and reduced incentives for selective migration is important here. The evolution of the 'push factor' of migration from eastern Europe will be as much a function of the expectations concerning the characteristics and the time horizon of the economic and social catching-up process as of the actually observed initial income gaps.

In the context of a potential process of catching up, specialization structures (reflected either in trade flows or in intra-corporate production location decisions) are symptomatic only of *specific phases of that catching-up process* and differ between the phases (see the by now voluminous literature on the South East Asian development process). As there is evidence already that catching up is and will continue to proceed at widely different speeds (if at all) in different regions of Central and Eastern Europe, the pattern of industrial specialization is differentiated both across the time dimension (following the phases of a development process) and the regional dimension. Hence the impact of East–West European integration and of industrial specialization upon labour markets in both western and eastern Europe should be seen in the context of this dynamic and differentiated process.

There is still great uncertainty about the longer-term characteristics of East–West European integration. The argument made here is that these characteristics depend particularly upon the extent, speed and nature of catching-up processes of CEE economies. Consequently, one should be careful in drawing conclusions from applications of relatively static theories of trade and industrial specialization to the process of East–West European integration; this care should extend to the analysis of the implications of this integration process for the structural dynamics of eastern and western European labour markets.

Looking at the integration process from a dynamic perspective, one should pay increasing attention to the potential impact of the economic integration process of these two complementary parts of the European continent upon the *dynamism of the European economy as a whole*. Concern for the analysis of the timeframe of catching-up processes in the current liberalized conditions of East–West European relations, of the factors which constrain the embarkation upon speedy catching-up including the extent, geographical coverage and characteristics of cross-border corporate linkages and of the emerging regional diversity of growth processes in this region should become paramount.

5 CONCLUSION

This chapter has analysed the prospects for convergence and integration between the EU member states and the applicant countries of CEE, and some of the implications of emerging trends for competitiveness. The analysis of these factors suggests that the EU will face a prolonged period of significant economic disparities between member states. The EU could take 10–15 years to take in all of the CEE-10 applicants, given the prospects for negotiating periods and EU reforms (see Grabbe and Hughes, 1998). The gap in income levels is likely to narrow somewhat during the period prior to accession, although this process is unlikely to be dramatic, given the outlook for growth rates discussed in section 2 and for catching-up processes in section 3.

In addition to the differences between new and old member states, the EU also has to consider the gap between the CEE countries, which might be widening during the period of accessions. Section 3 has shown the considerable variation across the CEE-10 in their integration with the EU through trade and investment. The richer CEE countries are most integrated with the EU in this respect, and the differences in FDI inflows between the top three recipients (Hungary, Poland and the Czech Republic) and the rest is striking. Because these three countries are the front-runners for EU membership at present (as well as the first new NATO members), accession prospects are reinforcing other trends in speeding their integration relative to the other CEE applicants. Moreover, earlier accession for the front-runner countries is likely to boost their integration further, and they will get more financial transfers from the EU than those left outside (see Grabbe and Hughes, 1998).

This unevenness in integration patterns across both the EU and CEE has implications for convergence. There is considerable heterogeneity in degrees of integration with CEE across the EU-15, and the impact of integration might be to enhance the competitiveness of the most integrated countries (especially Germany) *vis-à-vis* the less integrated countries and regions. In CEE, greater integration with the EU economy and earlier accession are likely to encourage faster catching-up in the front-runner countries in comparison with those that are currently further behind, opening up wider disparities between them.

APPENDIX 9.1 INTER-INDUSTRY TRADE SPECIALIZATION OF CEE PRODUCERS IN EU MARKETS

In this appendix we report on an exercise which estimates the factor contents of export flows from CEE to EU markets. The statistics used are *Eurostat* trade statistics (at the 3-digit NACE level) and the idea is to obtain a picture of factor intensity biases of CEE countries' (and comparative western and southern European countries') exports to EU markets.

Industries were ranked using EU production information concerning relative factor contents and then this information was superimposed to analyse the structure of exports of individual countries to EU markets. Five factors were distinguished:

Factor intensities	*Definition of statistical indicators*
Capital	Cumulative investment/number of employees
Labour	Number of employees/output
R&D	R&D expenditure/value added
Skill	Non-manual labour/total employment
Energy	Energy costs/output

The information concerning factor intensities for 3-digit NACE industries refers to values derived for the four largest EU member states (a proxy for the EU average); this information was previously used in *European Economy* (1995) and Landesmann (1995). With this information we can trace the relationship between a country's export specialization to EU markets and the factor content of these exports. Both the advantages as well as the disadvantages of adopting this procedure must be pointed out. The advantage of using production statistics solely derived from EU statistics is that patterns of trade are examined in relation to a common ranking of industries in terms of capital-, labour-, energy-, R&D- and skill-intensities. The drawback in adopting these measures is that individual countries' actual factor contents are very likely to deviate from this common measure. However, as long as the rankings of industries in terms of relative factor intensities do not differ very much across countries – even if the absolute levels do – the analysis should still give an insight into the relative factor contents of different countries' exports (for further discussion on this, see Landesmann, 1995).

Table 9A.1 presents the shares in total exports to the EU of the 10, 20, 30 most capital-, labour-, energy-, R&D- and skill-intensive industries (out of 92 3-digit manufacturing industries) in relation to the shares of these industries in total EU imports (excluding intra-EU trade).[3] Positive values mean a stronger (percentage) representation of these industries in a country's exports to the EU compared to the EU's total imports; negative values mean a (percentage) under-representation of such industries.

The general picture which emerges is that, in comparison to other (non-EU-12) exporters to the EU market, the exports from CEE economies are:

- strongly under-represented in skill- and R&D-intensive areas;
- more strongly represented in labour-intensive branches;
- less strongly represented (except for Slovakia) in capital-intensive branches; and
- except for Hungary and Slovenia, there is still an overhang of a strong representation in energy-intensive branches.

Against this broad picture, we can discern significant differences in the degree to which capital-, labour-, energy-, R&D-, and skill-intensive industries are under- or over-represented across the different CEE economies and in the trends since 1989. In particular, the under-representation of R&D- and skill-intensive industries is significantly lower by 1996 in Hungary, the Czech Republic, Slovenia and – to some extent – Slovakia, than in the other CEE economies; the positive trends of Hungary and the Czech Republic in this respect are particularly remarkable. There are also significant inter-country differences (for econometric estimates of these, see Landesmann, 1995) in the degree to which CEE exports have shifted towards labour-intensive and away from capital-intensive branches since 1989.

To sum up the analysis so far, we have evidence of significant inter-industry specialization structures in CEE's trade relationships to the EU. The specialization patterns have been significantly affected (partly strengthened, partly changed) by developments since 1989. A general 'developmental gap' (compared to general EU imports) in specialization biases of CEE economies away from capital-, R&D- and skill-intensive branches and towards labour-intensive exports can be discerned. However, the strength by which such specialization patterns were reinforced or modified since 1989 differed markedly across the different CEE economies; the more western CEE economies did not experience a reinforcement of this pattern of specialization and, in the cases of Slovenia, Hungary and the Czech Republic, we can perceive a significant closure of the gap in skill- and R&D-intensive areas; in the cases of Poland, Romania and Bulgaria the gap remains wide or is even widening. This process of differentiation reflects differences both in the developmental levels which reveal themselves in a climate of general trade liberalization, and in the dynamics of industrial developmental change that has taken place in the various CEE economies since 1989.

Table 9A.1 Over-/under-representation of the 10, 20, 30 most x-factor-intensive industries in the CEE export structure to the EU (difference from total EU import structure, in %)

	HUNGARY								POLAND							
	1989	1990	1991	1992	1993	1994	1995	1996	1989	1990	1991	1992	1993	1994	1995	1996
10 Capital	−11.61	−11.20	−10.68	−10.43	−9.95	−10.12	−10.02	−7.57	−10.50	−10.02	−8.90	−8.85	−8.06	−8.14	−8.65	−8.56
20	−14.87	−14.06	−16.87	−16.10	−14.34	−8.89	−1.91	−3.16	−3.66	−5.88	−5.90	−3.50	−4.32	−1.58	−2.97	−4.78
30	−17.54	−16.94	−18.58	−19.07	−17.92	−12.57	−5.28	−3.16	−2.51	−6.39	−7.70	−7.10	−8.28	−5.89	−6.70	−7.84
	1989	1990	1991	1992	1993	1994	1995	1996	1989	1990	1991	1992	1993	1994	1995	1996
10 Labour	10.29	9.81	9.28	11.17	9.94	7.47	3.59	3.73	6.60	6.90	8.82	9.67	10.15	8.89	7.67	6.72
20	8.93	9.25	8.67	10.20	10.00	6.82	2.68	2.23	5.73	5.84	7.36	7.92	10.71	8.65	7.57	7.22
30	9.56	9.79	9.20	11.31	11.45	9.39	4.91	2.23	3.75	4.77	6.38	7.37	10.16	8.61	7.27	8.04
	1989	1990	1991	1992	1993	1994	1995	1996	1989	1990	1991	1992	1993	1994	1995	1996
10 R&D	−16.05	−15.43	−15.67	−12.75	−10.33	−8.36	−7.17	−4.20	−18.95	−18.80	−21.63	−20.87	−19.02	−19.53	−18.73	−18.11
20	−20.33	−20.14	−19.66	−15.52	−11.53	−5.99	3.38	2.23	−19.60	−21.79	−25.49	−22.21	−17.56	−19.18	−17.64	−15.57
30	−26.71	−26.14	−24.05	−20.89	−16.42	−10.41	−1.16	2.23	−26.70	−28.99	−32.93	−30.70	−25.21	−26.05	−24.24	−22.03
	1989	1990	1991	1992	1993	1994	1995	1996	1989	1990	1991	1992	1993	1994	1995	1996
10 Skill	−13.74	−14.09	−12.87	−12.00	−10.99	−10.77	−10.35	−7.68	−14.87	−14.84	−14.27	−14.19	−13.84	−14.16	−14.34	−13.56
20	−21.39	−20.64	−19.49	−16.84	−13.64	−12.18	−11.80	−14.36	−23.88	−23.76	−25.32	−25.07	−23.70	−23.58	−22.92	−21.50
30	−28.07	−27.28	−25.55	−22.91	−19.93	−18.22	−17.74	−14.36	−30.03	−30.50	−32.76	−31.98	−28.41	−29.53	−28.55	−26.94
	1989	1990	1991	1992	1993	1994	1995	1996	1989	1990	1991	1992	1993	1994	1995	1996
10 Energy	−2.28	−0.64	−3.04	−3.12	−3.28	−1.30	−2.97	−3.33	5.91	5.73	7.44	7.71	3.22	6.94	4.95	3.00
20	−1.09	0.48	−2.33	−3.03	−3.35	−1.75	−3.47	−2.93	6.85	6.76	8.40	8.15	3.44	7.33	5.34	3.12
30	−1.91	0.62	−1.90	−1.86	−0.74	0.66	−2.14	−2.93	6.33	7.67	8.23	8.14	4.12	8.36	7.50	5.49

	CZECH REPUBLIC				SLOVAK REPUBLIC				SLOVENIA			
	1993	1994	1995	1996	1993	1994	1995	1996	1993	1994	1995	1996
Capital												
10	–5.05	–6.25	–7.09	–7.81	–1.73	–0.85	–1.85	–5.28	–8.44	–8.42	–7.83	–8.59
20	–3.68	–5.82	–8.17	–8.91	2.37	5.33	8.02	6.90	–5.73	–2.56	–1.12	0.35
30	–6.73	–9.63	–10.85	–11.67	–1.31	0.69	2.49	0.82	–13.00	–9.33	–8.03	–6.61
Labour												
10	1.68	1.79	–0.47	–1.09	5.75	4.32	2.79	3.60	6.21	4.00	1.95	2.24
20	0.86	1.82	–0.02	–1.03	5.76	5.55	3.10	2.63	6.78	3.42	1.61	0.83
30	3.02	4.54	3.45	3.80	2.61	3.69	0.81	0.57	8.74	5.55	3.77	3.74
R&D												
10	–15.76	–14.79	–13.26	–13.11	–20.44	–18.27	–18.06	–17.16	–11.11	–11.10	–10.99	–11.23
20	–13.52	–13.81	–11.25	–8.90	–23.30	–19.10	–13.18	–8.80	–4.31	–2.26	–1.26	0.45
30	–16.45	–15.55	–12.23	–9.19	–28.93	–22.92	–17.07	–13.82	–9.67	–7.60	–6.06	–5.10
Skill												
10	–8.21	–8.17	–7.96	–7.21	–9.67	–8.36	–8.60	–10.00	–8.07	–8.45	–8.51	–8.86
20	–17.63	–16.76	–15.40	–15.00	–20.34	–18.26	–18.03	–17.93	–11.26	–11.48	–11.63	–12.21
30	–21.95	–19.99	–17.56	–16.51	–22.88	–18.22	–19.08	–21.63	–16.56	–16.34	–15.19	–16.94
Energy												
10	6.07	5.15	3.06	1.83	16.23	15.95	12.82	11.27	–1.52	–0.56	0.23	0.36
20	9.58	8.31	6.74	5.06	20.73	20.24	15.92	14.12	0.16	0.99	1.19	0.97
30	10.22	9.68	8.92	7.74	20.97	20.53	17.43	17.31	1.94	2.65	2.54	2.36

Table 9A.1 continued

	ROMANIA								BULGARIA							
	1989	1990	1991	1992	1993	1994	1995	1996	1989	1990	1991	1992	1993	1994	1995	1996
10 Capital	–9.35	–8.74	–6.74	–8.95	–9.06	–9.97	–10.00	–10.20	–8.51	–9.87	–9.30	–8.85	–9.10	–9.80	–10.01	–10.23
20	–6.75	–11.18	–12.57	–8.95	–15.87	–7.79	–5.45	–10.22	–2.71	–1.44	–3.41	–6.24	–7.62	1.30	9.19	0.35
30	–13.17	–19.05	–20.19	–22.35	–23.92	–16.72	–14.49	–19.71	–2.74	–2.10	–3.49	–7.95	–10.29	–4.18	2.52	–5.82
	1989	1990	1991	1992	1993	1994	1995	1996	1989	1990	1991	1992	1993	1994	1995	1996
10 Labour	14.94	20.57	19.41	26.33	32.00	29.65	27.27	31.47	4.66	6.20	7.38	14.97	15.27	10.60	8.44	12.66
20	17.34	25.17	23.11	26.33	35.87	31.52	29.84	35.05	4.63	7.32	10.94	16.49	17.26	11.50	7.35	13.83
30	15.06	23.02	21.05	27.11	32.86	28.07	26.30	31.84	4.48	6.05	9.49	14.88	15.52	8.76	3.56	10.24
	1989	1990	1991	1992	1993	1994	1995	1996	1989	1990	1991	1992	1993	1994	1995	1996
10 R&D	–20.80	–20.61	–21.05	–21.59	–21.69	–21.33	–21.62	–23.09	–16.99	–18.92	–18.72	–18.98	–17.39	–17.22	–21.94	–22.84
20	–24.91	–25.80	–26.50	–21.59	–27.36	–27.24	–27.02	–28.28	–19.96	–23.23	–23.24	–23.23	–22.10	–22.23	–27.15	–28.38
30	–33.73	–34.27	–34.30	–35.83	–35.46	–35.27	–34.75	–35.50	–23.06	–27.80	–29.48	–30.50	–28.83	–29.61	–34.27	–34.95
	1989	1990	1991	1992	1993	1994	1995	1996	1989	1990	1991	1992	1993	1994	1995	1996
10 Skill	–14.78	–15.11	–15.05	–15.78	–15.19	–15.83	–16.09	–15.85	–7.14	–10.52	–10.63	–11.68	–10.87	–12.84	–14.33	–14.45
20	–26.71	–27.11	–26.95	–15.78	–27.29	–27.31	–27.90	–28.60	–17.55	–22.28	–21.45	–22.32	–21.07	–24.11	–26.68	–27.65
30	–33.75	–34.36	–34.21	–34.92	–33.95	–34.29	–34.27	–34.57	–19.83	–26.14	–24.04	–28.17	–26.84	–26.41	–32.67	–32.58
	1989	1990	1991	1992	1993	1994	1995	1996	1989	1990	1991	1992	1993	1994	1995	1996
10 Energy	7.54	4.57	3.45	1.58	–0.24	7.56	9.79	5.59	6.33	7.44	6.97	7.33	6.25	14.75	22.60	14.62
20	6.90	3.59	2.21	1.58	–0.93	6.50	9.50	5.68	7.80	8.38	6.82	7.68	5.96	13.79	21.79	13.32
30	4.96	1.44	–0.45	–1.87	–3.87	4.36	7.13	3.08	6.75	7.23	5.13	6.39	5.39	12.75	19.68	11.33

Source: Eurostat Trade Statistics; own calculations.

NOTES

1. Undervaluation here is simply defined in relation to the purchasing power value of a currency (PPP) and not in relation to targeting external accounts equilibria. The evidence suggests that there was a significant decline in the gaps between the PPP and the current exchange rates of CEECs since 1991 and also a convergence of the ratios between PPP and current exchange rates across CEECs (see Landesmann, 1998).
2. CEE-6 are Bulgaria, the Czech Republic, Hungary, Poland, Romania and Slovakia. The Baltic states and Slovenia are excluded from all 1989 totals because their trade was not recorded separately by the IMF prior to their independence.
3. Total EU imports here refers always to the imports of the EU-12 (that is, before the accession of Austria, Sweden and Finland) and excludes intra-EU-12 trade flows. This choice was made for consistency reasons.

REFERENCES

Baldwin, Richard E. (1994), *Towards an Integrated Europe*, London: Centre for Economic Policy Research.

Baldwin, Richard E. (1995a), 'The Eastern Enlargement of the European Union', *European Economic Review*, **39**, 474–81.

Baldwin, R.E. (1995b), 'The Effects of Trade and Foreign Direct Investment on Employment and Relative Wages', *NBER Working Paper*, No. 5037, also in *OECD Economic Studies*, No. 23, Winter 1993, 7–54.

Baldwin, Richard E., Joseph F. François and Richard Portes (1997), 'The Costs and Benefits of Eastern Enlargement', *Economic Policy*, **24** (April).

Borjas, G.J., R.B. Freeman and L.F. Katz (1992), 'On the Labour Market Effects of Immigration and Trade', in G.J. Borjas and R.B. Freeman (eds), *Immigration and the Work Force: Economic Consequences for the United States and Source Areas*, Chicago: University of Chicago Press, pp. 213–44.

Borrus, M. (1995), 'Left for Dead: Asian Production Networks and the Revival of U.S. Electronics', in E.M. Doherty (ed.), *Japanese Investment in Asia, International Production Strategies in a Rapidly Changing World*, San Francisco: The Asia Foundation, pp. 125–46.

Brown, Drusilla K., Alan V. Deardorff, Simeon D. Djankov and Robert M. Stern (1997), 'An Economic Assessment of the Integration of Czechoslovakia, Hungary and Poland into the European Union', in Stanley W. Black (ed.), *Europe's Economy Looks East: Implications for Germany and the European Union*, Cambridge: Cambridge University Press, pp. 23–63.

Cadot, Olivier, Riccardo Faini and Jaime de Melo (1995), 'Early Trade Patterns under the Europe Agreements: France, Germany and Italy', *European Economic Review*, **39**, 601–10.

CEPR (1990), *The Impact of Eastern Europe. Monitoring European Integration 1*, London: Centre for Economic Policy Research.

Collins, S.M. and D. Rodrik (1991), *Eastern Europe and the Soviet Union in the World Economy*, Washington, DC: Institute for International Economics.

Commission of the European Communities (1994), 'Trade Liberalisation with Central and Eastern Europe', *European Economy*, Supplement A (No. 7), 1–20.

Commission of the European Communities (1997), *Agenda 2000: For a Stronger and Wider Union*, Luxembourg: OOPEC.

Dittus, Peter and Palle S. Andersen (1995), 'Sectoral and Employment Effects of the Opening Up of Eastern Europe', in Patrick de Fontenay, Giorgio Gomel and Eduard Hochreiter (eds), *Western Europe in Transition: The Impact of the Opening Up of Eastern Europe and the Former Soviet Union*, Washington, DC: International Monetary Fund, pp. 55–90.

Doherty, E.M. (ed.) (1995), *Japanese Investment in Asia, International Production Strategies in a Rapidly Changing World*, San Francisco: The Asia Foundation.

Drábek, Zdenek and Alasdair Smith (1995), *Trade Performance and Trade Policy in Central and Eastern Europe*, Discussion Paper No. 1182, London: Centre for Economic Policy Research.

EBRD (1995), *Transition Report*, London: European Bank for Reconstruction and Development.

EBRD (1997), *Transition Report*, London: European Bank for Reconstruction and Development.

Estrin, Saul, Kirsty Hughes and Sarah Todd (1997), *Foreign Direct Investment in Central and Eastern Europe: Multinationals in Transition*, London: RIIA/Cassell.

European Economy (1995), 'The Interpenetration between the EU and Eastern Europe', *European Economy. Reports and Studies*, No. 6.

Eurostat, *External and Intra-EU Trade: Monthly Statistics*, Luxembourg: Statistical Office of the European Communities, various issues.

Faini, Riccardo and Richard Portes (1995), 'Opportunities Outweigh Adjustment: The Political Economy of Trade with Central and Eastern Europe', in Riccardo Faini and Richard Portes (eds), *European Union Trade with Eastern Europe: Adjustments and Opportunities*, London: Centre for Economic Policy Research, pp. 1–18.

Feenstra, R.C. and G.H. Hanson (1995a), 'Foreign Investment, Outsourcing and Relative Wages', *NBER Working Paper*, No. 5121, Cambridge, MA.

Feenstra, R.C. and G.H. Hanson (1995b), 'Foreign Direct Investment and Relative Wages: Evidence from Mexico's Maquiladoras', *NBER Working Paper*, No. 5122, Cambridge, MA.

Grabbe, Heather and Kirsty Hughes (1998), *Enlarging the EU Eastwards*, Chatham House Paper, London: The Royal Institute of International Affairs.

Graziani, G. (1997), 'The Relocation of the Italian Textiles and Clothing Industry to Eastern Europe: Outward Processing Traffic and Foreign Direct Investment'; ACE project paper; forthcoming in M. Landesmann and R. Dobrinsky (eds), *Corporate Restructuring and Economic Recovery in Central and Eastern Europe*, Aldershot: Edward Elgar.

Hallet, Martin (1997), *National and Regional Development in Central and Eastern Europe: Implications for EU Structural Assistance*, Economic Paper No. 120, March, Brussels: Commission of the European Communities.

Hamilton, Carl B. and L. Alan Winters (1992), 'Opening Up International Trade with Eastern Europe', *Economic Policy*, **14** (April), 77–116.

Hughes, Gordon and Paul Hare (1992), 'Trade Policy and Restructuring in Eastern Europe', in John Flemming and J.M. Rollo (eds), *Trade, Payments and Adjustment in Central and Eastern Europe*, London: Royal Institute of International Affairs, pp. 181–210.

Hughes, Kirsty (1996), 'European Enlargement, Competitiveness and Integration', in Pat Devine, Yannis Katsoulacos and Roger Sugden (eds), *Competitiveness, Subsidiarity and Industrial Policy*, London and New York: Routledge.

Hunya, G. (1997), 'Foreign Direct Investment in Transition Countries in 1995', *The Vienna Institute Monthly Report*, No. 1, 2–8.

IMF (1996), *Direction of Trade Statistics Yearbook*, Washington, DC: International Monetary Fund.

Landesmann, M. (1995), 'The Pattern of East–West European Integration: Catching Up or Falling Behind?', *WIIW Research Reports*, No. 212, Vienna; also in R. Dobrinsky and M. Landesmann (eds), *Transforming Economies and European Integration*, Aldershot: Edward Elgar.

Landesmann, Michael (1998), 'Features of East–West European Integration: Cost Structures and Patterns of Specialization', *EBRD Working Paper*, London.

Landesmann, M. and J. Poeschl (1997), 'Balance-of-Payments Structures and Macroeconomic Growth in Central and Eastern Europe (with J. Poeschl)', Ch. 7 in *Structural Developments in Central and Eastern Europe – WIIW Report*, Vienna.

Lawrence, R.Z. (1996), 'Single World, Divided Nations, International Trade and OECD Labour Markets', OECD Development Centre, Brookings Institution Press.

Leamer, E. (1994), 'Trade, Wages and Revolving Door Ideas', Working Paper 4716, Cambridge, MA: National Bureau of Economic Research.

Markusen, J.R., A.J. Venables, D.E. Konan and K.H. Zhang (1996), 'A Unified Treatment of Horizontal Direct Investment, Vertical Direct Investment, and the Pattern of Trade in Goods and Services', *NBER Working Paper*, No. 5696, Cambridge, MA.

Murphy, K.M. and F. Welch (1991), 'The Role of International Trade in Wage Differentials', in M. Kosters (ed.), *Workers and Their Wages: Changing Patterns in the United States*, Washington, DC: American Enterprise Institute.

Neven, Damien (1995), 'Trade Liberalization with Eastern Nations: How Sensitive?', in Riccardo Faini and Richard Portes (eds), *European Union Trade with Eastern Europe: Adjustments and Opportunities*, London: Centre for Economic Policy Research, pp. 19–60.

Rollo, Jim and Alasdair Smith (1993), 'The Political Economy of Eastern European Trade with the European Community: Why So Sensitive?', *Economic Policy*, **16** (April), 139–81.

Sachs, J.D. and A.M. Warner (1996), 'Achieving Rapid Growth in the Transition Economies of Central Europe', Harvard Institute for International Development, *Development Paper*, No. 544, July.

Sachs, J.D. and H.J. Shatz (1994), 'Trade and Jobs in U.S. Manufacturing', *Brookings Papers on Economic Activity*, **1**, 1–84.

UN (1997), *World Investment Report*, New York and Geneva: United Nations.

Urban, W. (1997), 'Patterns of Structural Change in Manufacturing in the Transition Countries', in M. Landesmann (ed.), *Structural Developments in Central and Eastern Europe*, Vienna: The Vienna Institute for International Economic Studies.

Wang, Z. and L.A. Winters (1991), 'The Trading Potential of Eastern Europe,' *CEPR Discussion Paper*, No. 610.

Winters, L. Alan (ed.) (1995), *Foundations of an Open Economy: Trade Laws and Institutions for Eastern Europe*, London: Centre for Economic Policy Research.

Wood, A. (1995), 'How Trade Hurt Unskilled Workers', *Journal of Economic Perspectives*, **9**, Summer, 57–80.

World Bank (1996), *World Development Report 1996: From Plan to Market*, New York: Oxford University Press.

Wyplosz, Charles (1995), 'Trade and European Labour Markets', in Patrick de Fontenay, Giorgio Gomel and Eduard Hochreiter (eds), *Western Europe in Transition: The Impact of the Opening Up of Eastern Europe and the Former Soviet Union*, Washington, DC: International Monetary Fund, pp. 55–90.

Zemplinerova, A. (1997), 'The Role of Foreign Enterprises in the Czech Economy', *The Vienna Institute Monthly Report*, No. 1, 18–24.

10. Conclusions and policy implications

Iain Begg, Jan Fagerberg and Paolo Guerrieri

Europe's performance relative to the USA and countries in Asia is a topic that greatly preoccupies policy-makers who are concerned that the EU is losing ground compared to other, more dynamic parts of the world. Although the recent crises in Asia gave a timely reminder that the grass often looks greener on the other side of the fence, the evidence assembled in this book points to trends in EU performance that policy-makers will find disconcerting.

In many areas, Europe has not kept pace with the technological advances of competitors and seems to have lost its dynamism. Employment creation has been lacklustre, new specializations have been slow to emerge and the rate of innovation has been disappointing. Redressing this relatively disappointing performance will be neither easy nor quick, but if enduring answers to Europe's problems are to be found, it is essential that the scale and nature of these be carefully diagnosed and solutions found.

This concluding chapter aims to contribute to this debate. It summarizes the key messages from the preceding chapters and draws out implications for policy. The next section explores the policy challenges confronting the EU. This is followed by a summary of the key findings and evidence presented in this book. The chapter concludes by drawing out lessons for policy.

1 THE CHALLENGES TO POLICY

Although criticisms can reasonably be levelled at the conduct of macro-economic policy in the last quarter of a century, the main purpose of this book has been to explore the deficiencies of the supply side of the European economy. Several of the chapters document in detail the shortcomings of the EU's performance in areas such as economic growth, internal convergence, competitiveness and employment. Productivity growth has slowed down relative to competitors (Chapter 1). While until recently there was a tendency towards convergence in productivity and income between European regions, there are now signs of a reversal of this trend (Chapter 6). Export competitiveness has deteriorated in all areas except agriculture and raw materials (Chapters 1 and 2). The losses have been most manifest in the technologically most sophisticated industries, particularly for ICT products (Chapters 2 and 5). Europe has also failed to create employment on a scale at all comparable with the USA or Japan, with obvious repercussions for unemployment (Chapter 4).

The challenge for policy-makers is no less than devising solutions which restore the dynamism and creativity that characterized the European economy in earlier periods (Chapter 1). Innovation and the application of new technologies have been shown to be central to the explanation of Europe's relative economic performance. As was noted in the introduction, however, the main focus of supply-side policy has, hitherto, been on making markets larger and more competitive. The approach here, by emphasizing the role of innovation and technology in shaping economic performance, suggests that fresh thinking is now needed on how to enhance the supply side.

It is imperative to achieve this increased dynamism without compromising the goal of a more egalitarian distribution of income across Europe. As was shown in Chapter 6, regional disparities in economic performance remain substantial, and have increased in many member countries in the last decade. This gap is especially marked as regards innovative activity. Hence, there is a risk that a faster rate of innovation, which is vital for European growth and competitiveness in general, might further aggravate regional disparities. The challenge for policy-makers is, therefore, to design policies that increase both innovation and the ability of the least favoured regions to absorb new technologies. Similar issues arise in relation to the eastern enlargement anticipated in the next few years. As documented in Chapter 9, the income and productivity levels of the potential new entrants are low compared to the EU, with a great deal of internal variation. If, following accession, the CEECs are to avoid falling further behind, it will be incumbent on the EU to contribute to enhancing their technological capabilities.

Perhaps the most important problem facing EU policy-makers is how to combat the persistent high rate of unemployment in Europe, which threatens social cohesion. It is important to stress that innovation is, in itself, no 'quick fix' for unemployment. Product innovation may increase employment through increased demand for products embodying the new technologies. But process innovation aimed primarily at rationalization of existing production processes may also reduce employment in a specific industry and/or location, if not compensated by indirect income and demand effects. It is shown in Chapter 4 that the latter (employment-reducing) type of outcome is the most common in Europe, in sharp contrast to the situation in the other parts of the 'triad'. Thus the challenge for policy-makers is not only to increase innovation and diffusion, but to do so in a way that is more 'employment friendly'.

2 DEFINING THE AGENDA: PRINCIPAL FINDINGS

The various contributions to this book provide a careful analysis of the strengths and weaknesses of Europe and highlight key policy issues that need to be addressed.

2.1 Specialization, Competitiveness and Growth

Although there are differing views on what form of specialization is most desirable for a country, it is hard to dispute the view that specialization matters. But how? Traditional economic theory tells us that specialization is beneficial in itself because it leads to more efficient use of available resources. However, unless this has a positive effect on technological progress ('dynamic economies of scale'), it will not lead to higher growth in the long run. The empirical evidence considered in Chapter 2 confirms this by demonstrating that increasing the degree of specialization (irrespective of activity) is not a foolproof recipe for achieving higher growth in the long run.

Rather, the analysis shows that what matters for growth and competitiveness is not so much increasing the degree of specialization in general, as the ability to exploit areas of high technological opportunity, which in recent years have been dominated by ICT (Chapter 5). However, the analysis shows that Europe has lost ground in a number of strategically important sectors, particularly those related to ICT. This raises the issue of what policy-makers can and should do to reverse this uncomfortable trend. Chapter 2 argues that market mechanisms alone will not secure change. Instead, a long-run strategy for competitiveness has to be based on improving access to knowledge, enhancing the skills and adaptability of workers, and facilitating the diffusion of innovations. The aim should be to achieve a 'positive' flexibility through investment in productive resources rather than by short-term pressures on working conditions that ultimately prove to be counter-productive.

Traditionally, competitiveness has been identified with labour costs, although it is evident that Europe can never hope to compete with newly industrialized countries in Asia and elsewhere on the basis of labour costs alone. Moreover, the long-run outlook for Europe is also influenced by the quality of what it produces, increasingly so in a global market-place in which competition on non-price factors is becoming ever more intense. The evidence in Chapter 3 shows that although quality is to some extent linked to the level of economic development, a country's R&D effort and the level of skills in its labour force are also important influences. Together these factors go a long way in explaining the differences between EU member states in quality of exports. The chapter shows that there is great scope for raising quality as the core competitive advantage of the EU through a strategy based on higher product innovation, upgrading of skills and increased R&D efforts.

The competitiveness of EU firms, in the EU market as well as in a global context, will also be affected by the challenge from the eastern enlargement anticipated in the next few years (Chapter 9). The research shows that the applicant countries mainly produce low-cost, low-quality products of a type that the EU already mostly imports, so that the consequences for the EU industry as a whole are likely to be limited. For the low-cost, low-quality segments of West European industry, however, the increased competition from CEE producers is likely to exert significant pressure on costs for quite some time. This will put additional pressure on quality upgrading in industries and areas exposed to such competition, and hence increase the danger of increased unemployment and declining social cohesion. This clearly underlines the importance of policies that put the prime focus on upgrading quality, skills and technological capabilities in affected industries and areas.

It is argued in Chapter 9 that the main competitive advantage of at least some of the new applicants is not so much cheap labour as cheap skilled labour. In the longer term this may lead to structural changes in their production and exports towards higher quality. The finding in Chapter 3 that the 'western' CEECs – particularly Hungary, the Czech Republic and Slovenia – were rapidly repositioning their exports in this direction, seems to confirm this view. If this trend continues, it should be welcomed insofar as it facilitates more rapid 'catch-up' and leads to increased diversity and dynamism in European industry, although there is a risk that the outcome could be to widen disparities between the CEECs.

2.2 Technology and Unemployment

With unemployment in the EU seemingly stuck at around 18 million, attention has increasingly been focused on the apparent inability of the EU economy, unlike other 'global regions', consistently to generate enough jobs. Chapter 4 documents this facet of Europe's comparative economic performance and presents findings that can help to inform a policy response. The contrast between Europe and its two main 'global' competitors is striking. In terms of number of jobs created, Europe lags far behind and the research also shows that the relationship between economic growth and employment creation is least favourable in Europe. This discrepancy means that Europe's industrial structure is orientated towards sectors most open to labour saving, notably mature industries characterized by a high degree of process innovations.

The challenge Europe faces is how to secure the growth of science-based and other high-growth industries and services, where product innovations dominate and markets are expanding. It is argued in Chapter 4 that too much weight has been given to a narrow competitiveness agenda that bears down on labour costs without paying enough attention to the longer-term aim of securing structural change towards a more 'favourable' specialization pattern. In parallel, policy-makers appear to have neglected the employment intensity of growth in favour of stability and cost-cutting objectives, and coherence between demand and supply policies has been lacking.

A major policy issue here is the extent to which it is desirable to target the expansion of sectors which offer good potential for growth, especially if these sectors are 'employment-friendly'. Past policy errors have resulted in such targeting falling out of favour. However, one of the threads running through this book is that a new technological paradigm based on ICT is emerging, and in the light of this it makes sense to look afresh at the policies that are available in this area.

2.3 The ICT Revolution and Europe

The work reported in Chapter 5 reaffirms the message emerging from the analyses of trade, output and employment about the gap between Europe and other advanced economies. The data presented indicate that most European firms have not kept pace with their American and Asian competitors in the markets for ICT hardware, software and services. Indeed, Europe has fallen behind Japan as the second most important supplier of hardware after the USA, and is increasingly vulnerable in software and services.

Plainly, there is no single explanation for this apparent decline in competitiveness. Governments have been criticized for over-burdening companies with regulation and social charges; companies have, arguably, been slow to seize opportunities in the field and the combination of industrial and technological policy may have been inappropriately targeted. Fragmented national efforts to promote national champions or state aid for ailing companies, for example, may have stifled the development of new and dynamic companies, and it is often argued that Europe has an insufficiently developed venture capital industry compared with the USA. Furthermore, the potential for making use of ICT has been inhibited by the slow adaptation of structures, management systems and regulation. However, as Chapter 5 demonstrates, there are important differences within Europe in this respect, with a much more widespread use of ICT in some of the northern countries than their counterparts in the south.

Responding to ICT as a new techno-economic paradigm calls for an assessment of the factors that influence capabilities in this area. The work in Chapter 5 stresses the importance of transforming the education and training systems in order to equip individuals with the skills needed for an environment in which the major new technology is pervasive. The continuing skill shortages in software testify to the failure of Europe to meet this challenge, and it is evident that this deficiency has slowed the diffusion of ICT beyond the immediate sectors that developed and applied these technologies. Chapter 5 also emphasizes the importance of combining skills with diffusion-oriented policies centred on social needs, in order to stimulate learning processes. Mobile communications in the Nordic area is a good example of how public regulation and support managed to bring together new technology, skills and existing social needs in a way that led to both rapid diffusion of the new technology and – through learning – the development of a globally competitive industry. The possibility that similar policies also may work for other types of ICT applications should encourage policy-makers to experiment further with diffusion-oriented policies centred on users and social needs.

2.4 Innovation Systems and Regional Dynamics

Economic and social cohesion – usually defined in terms of equity considerations such as regional disparities or social inclusion – is a fundamental aim of the EU articulated in Article 2 of the Treaty. Regional development is, however, also a factor in the overall performance of any economy. If there are too many regions in which productivity lags and which lack dynamism, the scope for boosting the supply side of the overall economy will be constrained. Thus, although the convergence achieved at member state level in the EU is encouraging, the evidence presented in Chapter 6 about persistent – and to some extent increasing – differences within countries in economic performance is a cause for concern.

In seeking to explain why the less-favoured regions have performed so disappointingly, the chapter focuses on three dimensions: technology, structure and institutions. Although data are scarce on many factors of potential relevance for regional growth, the evidence clearly indicates that most low-income regions have failed to exploit the potential for technology diffusion. This points to a need for policies aimed at enhancing the capacity of such regions to absorb new technologies, especially ICTs. To a large extent, the low rate of diffusion is associated with a structure of activity dominated by agriculture or 'older' industries, and a corresponding lack of high-tech activities, often combined with relatively high unemployment. One strategy for change would be to reorient policies in this area towards accelerating structural change. But it is important to design policies in a way that does not lead to a further increase in long-term unemployment, as this is itself a factor hampering diffusion and growth.

Regional policy-makers have learned the hard way that attempts to apply a single formula in different circumstances are doomed, because of different institutional structures, diverse starting conditions and great variations in internal dynamics. This is confirmed in the empirical analysis in Chapter 6. Further evidence comes from the detailed investigation in Chapter 8 of the North Jutland and Western Sweden regions, which shows the importance of looking at local conditions in understanding why and how particular 'clusters' of economic activity emerge. The research also drew attention to the potential for the local science base, in the form of university research, to contribute to local learning processes and, hence, regional development. While concepts such as the 'learning region' have become increasingly prominent in the regional policy literature, the translation of these into specific policy actions is at an early stage. Nevertheless, finding ways of boosting regional innovation and thus contributing to the enhancement of innovation and diffusion of new technology, should be high on the policy agenda.

Part of the answer has to do with getting a proper understanding of the 'system of innovation'; the mix of characteristics, infrastructure and policies that determines how well an economy is able to exploit opportunities afforded by new technologies. The research reported in Chapter 7 shows the wide variations between EU regions in patenting activity, a variable closely linked to invention and innovation. Patenting is concentrated in relatively few regions and there is evidence of clustering of such regions in a few 'hot-spots' across the EU.

Differences in invention and innovation would not matter much if knowledge spread readily from region to region. However, the research reveals that the greater the geographical distance between regions, the lower the degree of knowledge flows between them. Moreover, knowledge flows are greater within countries than between them, suggesting that the national element in innovation systems remains strong. The chapter also shows that knowledge flows are most intense between regions with similar or complementary specialization patterns, as well as between technologically linked sectors.

These results suggest powerful influences leading to the formation of geographically concentrated clusters of technologically related activities at work in Europe's economy. This clearly shows the interrelatedness of the many factors that contribute to regional technological capability. The findings also emphasize that fostering strong innovative clusters at the regional level is no easy task, and requires the coordination of many traditionally distinct policy areas. One of the more important challenges for the EU is to persuade policy-makers in different areas and at different levels to work together in a joint effort to boost regional dynamism.

3 LESSONS

The core message from this book is that the problems that Europe faces in key areas such as growth, equality and employment are all related to its failure to take sufficient advantage of technological advances, particularly the ICT revolution. Consequently, a coherent European strategy for upgrading technological capability and quality competitiveness is long overdue. This cannot be limited to providing support to selected industries (or companies) in order to make them more competitive in global markets. Rather, what Europe has to do is to take steps to embed new technologies, especially ICTs, in society. This should bring together regulation, science and technology policy, and employment initiatives. It is also important to ensure that macroeconomic policy complements rather than conflicts with supply-side policy, especially in the post-EMU environment when much will change. Equally, it is vital at the outset to recognize that change will not happen overnight and that boosting the long-run ability of the economy to create and use new technologies will require concerted action. The complementarities between policy areas, in particular, should also be stressed.

To some extent Europe's current problems are the price to be paid for past successes. European integration, most recently through the single market programme, has made product and factor markets more open and paved the way for the realization of economies of scale and a more efficient allocation of resources. From a long-run perspective, these policies have been extremely successful. Without them, it is doubtful whether Europe would have managed to catch up with the USA to the extent that it did. However, the rewards for catch-up in capital- and scale-intensive sectors producing for the mass markets have been cashed in long ago, and now the rules of the game have changed. As shown in this book, science-based industries, especially those drawing heavily on ICT, have become the main driver of technological change and economic growth since the early 1980s. Although the ICT revolution has been under way for a long time, its major effects are just beginning to be felt and it is clear that Europe's performance in ICT, however measured, is far from satisfactory. It is our contention that a mere continuation of past policies will not suffice to reverse this discomforting trend. If appropriate steps are not taken now, it is our assessment that the current trends toward slow growth, increasing inequality and unemployment will continue, threatening social cohesion, the social model and the democratic values that Europeans hold dear.

The story is easily told. Slow growth is mainly the result of failure to exploit the technological opportunities inherent in new and fast-growing technologies. This is not mainly a question of failing to be competitive in, say, the production of computers or other products embodying ICT, but of embedding the new technology in society at large. Regional disparities have been exacerbated by the very uneven diffusion of new technology, which has, hitherto, disproportionately favoured high-income, high-R&D regions specialized in high-tech manufacturing and – above all – services. By contrast, traditional agricultural regions or regions in the rustbelt have benefited very little if at all. Rather what has happened is that the relative prices of their products and factor services have declined. High unemployment is the flip side of this coin. Europe's continuing specialization in increasingly mature industries characterized by labour-saving innovation has meant that relatively few new jobs have been created. Low labour mobility has made matters worse by inhibiting effective redeployment of resources and has, arguably, been an impediment to technological advance.

Although the diagnosis is clear, agreeing on remedies is much less easy. What makes the problem so challenging is the need for actors and trends at very different levels to pull together. On the one hand, as amply documented in this book, the creation of technological advantage is in most cases a very local affair. At the other extreme, radical technological changes, particularly the ICT revolution, affect nearly all aspects of life, so that a holistic approach to change aimed at exploiting the opportunities afforded by these new technologies is needed. Because of path dependency in local technological, institutional and economic systems, and the complex conditions for getting the most out of ICT, this is very difficult to realize in practice. There is a large coordination problem here, and this may be why the more decentralized US system, and the much more authoritarian Japanese system, have both proved more effective in securing change than the multi-layered decision system in the EU.

Hence, our analysis points to systematic coordination failures and a need for a strategic view on European growth. The 'negative' integration policies (focusing on product and factor market integration) of the past have to be supplemented by new 'positive' policies on the supply side that enhance innovation; accelerate the diffusion, adoption and implementation of new technology; and improve the skills base of the labour force. Arguably, a much greater emphasis needs to be placed on stimulating more rapid, pervasive and effective innovation in technologies, organizations and institutions. These are themes that surface repeatedly in EU policy discourse but there is evident reluctance on the part of member states to agree to collective action at the EU level. This is nothing new, as there has long been a tension between member states and the Commission about competence over policy areas (education and training, for example, is one area in which any encroachment by the EU has been strongly resisted). However, fresh thought on this is required if the discomforting trends of recent decades are to be reversed. What is needed is a much more prominent role for the EU, not only in policy coordination, but in discussing policy change, experimenting with policy alternatives, evaluating the results and providing benchmarking for government at various levels.

A precedent can be found in some of the applications of the Structural Funds for regional development. Many of these are still being piloted or have only been tried in a limited number of regions. As they have evolved, the Structural

Funds have gradually devoted more effort to different sorts of technology initiatives. We advocate a progressive switch in the emphasis in the Funds towards more technology-orientated investments, and investments that facilitate structural change. It is also necessary to coordinate such policies with other policies at the EU or national level, such as the Common Agricultural Policy, so that regions face a consistent set of incentives (see Chapter 6). Structural change is bound to be more difficult to achieve if other policies at the EU or national level pull in the opposite direction.

We consequently maintain that Europe needs to adopt a wider and more inclusive definition of science and technology policy than in the past. EU-sponsored Framework Programmes have played a useful part in financing and encouraging specific scientific research. Little effort has, however, gone into promoting capability amongst 'users' of new technologies and in providing a social and institutional setting that encourages their implementation. The analysis in Chapter 5 suggests that this is a serious weakness and that a significant effort is needed to redress the position. In the past, the EU role has been rather limited, with exchanges of information and networking as examples of input. We argue that although the political imperative of subsidiarity and the need to take account of local regional and national diversity must receive due recognition, there is room for a more active EU policy in this area.

The work in this volume shows the scale and scope of the challenges that Europe faces in adjusting to globalization and advances in technology. We do not pretend that transformation of the supply side of the European economy will be easy or that the process will not be painful. But if unemployment is to be reduced and Europe's competitive edge restored imaginative new approaches to policy will be needed. The challenge to policy-makers is, therefore, to find the solutions that will allow these new approaches to be implemented.

Index